Advances in Intelligent Systems and Computing

Volume 1219

The series "Advances in Intelligent Systems and Computing" contains publications on theory, applications, and design methods of Intelligent Systems and Intelligent Computing. Virtually all disciplines such as engineering, natural sciences, computer and information science, ICT, economics, business, e-commerce, environment, healthcare, life science are covered. The list of topics spans all the areas of modern intelligent systems and computing such as: computational intelligence, soft computing including neural networks, fuzzy systems, evolutionary computing and the fusion of these paradigms, social intelligence, ambient intelligence, computational neuroscience, artificial life, virtual worlds and society, cognitive science and systems, Perception and Vision, DNA and immune based systems, self-organizing and adaptive systems, e-Learning and teaching, human-centered and human-centric computing, recommender systems, intelligent control, robotics and mechatronics including human-machine teaming, knowledge-based paradigms, learning paradigms, machine ethics, intelligent data analysis, knowledge management, intelligent agents, intelligent decision making and support, intelligent network security, trust management, interactive entertainment, Web intelligence and multimedia.

The publications within "Advances in Intelligent Systems and Computing" are primarily proceedings of important conferences, symposia and congresses. They cover significant recent developments in the field, both of a foundational and applicable character. An important characteristic feature of the series is the short publication time and world-wide distribution. This permits a rapid and broad dissemination of research results.

**** Indexing: The books of this series are submitted to ISI Proceedings, EI-Compendex, DBLP, SCOPUS, Google Scholar and Springerlink ****

More information about this series at http://www.springer.com/series/11156

Isabella Corradini · Enrico Nardelli ·
Tareq Ahram
Editors

Advances in Human Factors in Cybersecurity

AHFE 2020 Virtual Conference on Human Factors in Cybersecurity, July 16–20, 2020, USA

Springer

Editors
Isabella Corradini
Themis Research Center
Rome, Roma, Italy

Enrico Nardelli
Dip. Matematica
University of Rome Tor Vergata
Rome, Italy

Tareq Ahram
Institute for Advanced Systems Engineering
University of Central Florida
Orlando, FL, USA

ISSN 2194-5357 ISSN 2194-5365 (electronic)
Advances in Intelligent Systems and Computing
ISBN 978-3-030-52580-4 ISBN 978-3-030-52581-1 (eBook)
https://doi.org/10.1007/978-3-030-52581-1

This Springer imprint is published by the registered company Springer Nature Switzerland AG
The registered company address is: Gewerbestrasse 11, 6330 Cham, Switzerland

Advances in Human Factors and Ergonomics 2020

AHFE 2020 Series Editors

Tareq Z. Ahram, Florida, USA
Waldemar Karwowski, Florida, USA

11th International Conference on Applied Human Factors and Ergonomics and the Affiliated Conferences

AHFE 2020 Virtual Conference on Human Factors in Cybersecurity, July 16–20, 2020, USA

Advances in Neuroergonomics and Cognitive Engineering	Hasan Ayaz and Umer Asgher
Advances in Industrial Design	Giuseppe Di Bucchianico, Cliff Sungsoo Shin, Scott Shim, Shuichi Fukuda, Gianni Montagna and Cristina Carvalho
Advances in Ergonomics in Design	Francisco Rebelo and Marcelo Soares
Advances in Safety Management and Human Performance	Pedro M. Arezes and Ronald L. Boring
Advances in Human Factors and Ergonomics in Healthcare and Medical Devices	Jay Kalra and Nancy J. Lightner
Advances in Simulation and Digital Human Modeling	Daniel N Cassenti, Sofia Scataglini, Sudhakar L. Rajulu and Julia L. Wright
Advances in Human Factors and Systems Interaction	Isabel L. Nunes
Advances in the Human Side of Service Engineering	Jim Spohrer and Christine Leitner
Advances in Human Factors, Business Management and Leadership	Jussi Ilari Kantola, Salman Nazir and Vesa Salminen
Advances in Human Factors in Robots, Drones and Unmanned Systems	Matteo Zallio
Advances in Human Factors in Cybersecurity	Isabella Corradini, Enrico Nardelli and Tareq Ahram

(continued)

(continued)

Advances in Human Factors in Training, Education, and Learning Sciences	Salman Nazir, Tareq Ahram and Waldemar Karwowski
Advances in Human Aspects of Transportation	Neville Stanton
Advances in Artificial Intelligence, Software and Systems Engineering	Tareq Ahram
Advances in Human Factors in Architecture, Sustainable Urban Planning and Infrastructure	Jerzy Charytonowicz
Advances in Physical, Social & Occupational Ergonomics	Waldemar Karwowski, Ravindra S. Goonetilleke, Shuping Xiong, Richard H.M. Goossens and Atsuo Murata
Advances in Manufacturing, Production Management and Process Control	Beata Mrugalska, Stefan Trzcielinski, Waldemar Karwowski, Massimo Di Nicolantonio and Emilio Rossi
Advances in Usability, User Experience, Wearable and Assistive Technology	Tareq Ahram and Christianne Falcão
Advances in Creativity, Innovation, Entrepreneurship and Communication of Design	Evangelos Markopoulos, Ravindra S. Goonetilleke, Amic G. Ho and Yan Luximon

Preface

This book deals with role of the human factors in cybersecurity. It is in fact the human element what makes the cyberspace the complex and adaptive system it is.

According to international cybersecurity reports, people are both an essential part of the cybersecurity problem and of its solution. Understanding how people behave in the digital environment and the role of human error in successful security attacks is therefore fundamental for developing an effective approach to cybersecurity.

Cyberintrusions and attacks have increased dramatically over the last decade, exposing sensitive personal and business information, disrupting critical operations, and imposing high costs on the economy.

This book gathers studies on the social, economic, and behavioral aspects of the cyberspace and reports on technical and analytical tools for increasing cybersecurity. It describes new education and training methods for management and employees aimed at **raising** cybersecurity awareness. It **discusses** key psychological and organizational factors influencing cybersecurity.

Gathering the proceedings of the AHFE Conference on Human Factors in Cybersecurity, held virtually on July 16–20, 2020, this book offers a comprehensive perspective on ways to manage cybersecurity risks for a range of different organizations and individuals, presenting inclusive, multidisciplinary and integrated approaches combining the technical and behavioral elements.

Contributions have been organized into four sections:

Section 1 Cognitive Factors, Personality and Decisions Making
Section 2 Cybersecurity Tools and Analytics
Section 3 Awareness, Training and Education
Section 4 Social, Economic and Behavioral Aspects of Cybersecurity

Each section contains research papers that have been reviewed by members of the International Editorial Board. Our sincere thanks and appreciation to the board members as listed below:

July 2020

Isabella Corradini
Enrico Nardelli
Tareq Ahram

Contents

Cognitive Factors, Personality and Decisions Making

Creative Manual Code Obfuscation as a Countermeasure Against Software Reverse Engineering

Salsabil Hamadache[1,2](✉) and Malte Elson[1,2]

[1] Psychology of Human Technology Interaction Group, Faculty of Psychology, Ruhr University Bochum, Universitaetsstrasse 150, 44801 Bochum, Germany
{salsabil.hamadache,malte.elson}@rub.de

[2] Horst Görtz Institute for IT Security, Ruhr University Bochum, Universitaetsstrasse 150, 44801 Bochum, Germany

Abstract. Due to the relevance of IT security to industry, politics, and the public alike, research on IT-security-related issues is abundant. However, a lack of interdisciplinarity in this domain has led to a vast amount of detailed information on technical aspects of security one the one hand, and little to no insight into the psychological aspects of attacking, defending, or securely using technological systems on the other. This research effort aims to contribute to filling this gap by determining cognitive predictors of software reverse engineering as well as code obfuscation success and by describing and analyzing approaches and strategies IT specialists use when attacking or defending Java programs. Moreover, the relevance of adversarial reasoning in this domain is assessed. In an experimental design, participant pairs either receive an instruction into game theoretical concepts of adversarial reasoning or not, to then obfuscate Java code or reverse engineer clear and obfuscated code.

Keywords: Reverse engineering · Human factors · Code obfuscation · Problem solving · Adversarial reasoning

1 Introduction

When protecting software from adversarial attacks, considering how attackers and users will interact with it is paramount. Beyond technical considerations, a thorough understanding of cognitive approaches and strategies when developing, protecting, attacking, and using software products can further enhance IT security: Only when developers know who might attack their system and how they approach this undertaking, can they develop effective defenses. Instead of using heuristic or trivial defense practices which experienced adversaries likely will overcome, concrete anticipation of attacker behaviour may lead to greater security.

Attackers might want to gain access to and reverse engineer (RE) software code in order to find vulnerabilities in it, steal or illegitimately use intellectual property, or

I. Corradini et al. (Eds.): AHFE 2020, AISC 1219, pp. 3–8, 2020.
https://doi.org/10.1007/978-3-030-52581-1_1

create illegal copies of it. To prevent this from happening, software developers employ code obfuscation (CO) techniques that change code appearance while preserving its function. This might confuse attackers and thereby increase reverse engineering effort, thus making them reconsider the strength of their motivation and decide whether it is really worthwhile investing the required time and effort.

From a psychological perspective, both reverse engineering and code obfuscation can be conceptualized as problem solving processes [1], as complex problem solving has been defined as "a collection of self-regulated psychological processes and activities necessary in dynamic environments to achieve ill-defined goals that cannot be reached by routine actions" [2] and all of these elements can be found when observing reverse engineers at work: Even experienced reverse engineers would not consider their work to consist of routine actions and often report high effort and ever-changing circumstances. Goals are often not trivially defined but often rather consist of exploration of the program's functional logic. The alternation of attacking and defending code shows how dynamic the process can be.

Problem solving is often regarded in conjunction with creativity. We propose that code obfuscation, despite traditionally being a process performed by automatic tools, can also be conceptualized as a creative process of using one's own strengths to explore ways in which one can transform code aiming to make it more difficult to attack.

2 Past Research

Comprehension of (Obfuscated) Code. The investigation of code comprehension of regular (non-obfuscated) code, particularly within the context of answering the question of why programming seems to be such a difficult task, has extensively been studied for decades (e.g. [3]). Recently, publications in the field of code obfuscation have also been emerging (e.g. [4] and [5]). However, studies assessing the effectiveness of code obfuscation in experiments, using human participants, have been rare because both obfuscation and deobfuscation are often automatic, algorithmic processes. This stage has been set in 2014 by Ceccato and others [6] and joined in 2018 by Hänsch and colleagues [7]. In both studies, computer science students solved simple reverse engineering problems on two small Java programs, whereby one was presented to them in obfuscated code and the other was not. In both studies, the obfuscation method *opaque predicates*, in which unnecessary nesting and useless lines of code are added to the program code, was not effective in slowing down or preventing reverse engineering behaviour, whereas the obfuscation method *identifier renaming* was successful in doing so. This can be explained by the fact that students (or beginners) in particular, heavily rely on semantic information such as variable names when trying to make sense of code. Hänsch and colleagues further investigated the role of experience and found that experienced programmers are considerably better at reverse engineering clear code than beginners, but not at reverse engineering obfuscated code.

Adversarial Reasoning in IT Security. Hamman and colleagues [8] designed an experiment to demonstrate that adversarial reasoning is a highly valuable tool when defending assets in cybersecurity. In their study, participants either participated in a course on game theory concepts related to adversarial reasoning and their application in cybersecurity contexts or did not receive such a course. Then, they were asked to allocate defense resources depending on the likelihood and severity of an attack. Participants who had been educated on adversarial reasoning made better choices than the control group. It remains to be shown whether exposure to these concepts will also lead to higher performance in other, less game-theoretic cybersecurity-themed tasks.

Creativity and Intelligence in Hacking. For the purpose of talent detection as well as training development, it is – in any professional domain – sensible to investigate which cognitive variables are most relevant within given tasks. Often, intelligence is the main, but far from the only predictor of success in professional problem solving [2]. When solving complex problems, it might be more helpful to actively switch between so-called divergent and convergent processes [9], whereby the latter can be proxied by intelligence for simplification. Divergent thinking, however, relates to the ability to generate many different adequate ideas and think "out of the box" when trying to solve a problem. Research has shown that creativity is a core component of hacking [10], of which reverse engineering is one kind. Thus, an open research question is whether intelligence or creativity are more important to the reverse engineering and code obfuscation processes as well as to what degree they predict performance in tasks that relate to software protection and the vanquishing of them. It has also been shown that ambiguity tolerance is needed to be a successful hacker.

3 Method and Materials

Study Design. In a study with a yoked design, participants with basic knowledge of Java were randomly matched into code obfuscator (CO)/reverse engineer (RE) pairs, whereby the reverse engineers work on the code previously obfuscated by their study peers. This way, the better reverse engineers perform when working on the obfuscated code, the worse the quality of the code obfuscations performed by their partner. Code obfuscators receive:

- 0 points if they wreck the program;
- 1 point if their code remains functional, but their partner manages to quickly solve all reverse engineering tasks nonetheless;
- 2 points if their obfuscation at least slows the reverse engineer down, i.e. they are significantly slower on obfuscated than on clear code yet manage to solve all tasks;
- 3 points, if the reverse engineer manages to solve both tasks on clear code, but only one on obfuscated code, or one versus no task, respectively;
- 4 points, if the reverse engineer cannot solve a single task on obfuscated code despite being able to solve both tasks on clear code.

Participant pairs were randomly allocated to receive a game theory course prior to the code tasks, or no such training. For the RE group, their performance on obfuscated code is compared to their own performance on non-obfuscated code to investigate how obfuscation affects the RE process and to compare strategies and approaches applied on clear vs. obfuscated code. Thus far, mainly IT security and computer science students, as well as a few professional reverse engineers and students of other fields are also within our preliminary sample (N = 22).

Procedure. Participants report on their ambiguity tolerance, programming and reverse engineering experience, and basic demographics by answering questionnaires. After that, they are introduced to two simple Java programs, a racing game and a Chat Application. Participants in the CO group are then given one hour to perform any operation they come up with in order to make the code more complex for a human attacker, while leaving the code function intact. Participants in the RE group receive 4 questions inquiring where in the code specific information can be found or where specific program elements are implemented. They are asked to work as fast as possible within a time frame of 60 min. Lastly, participants perform creativity and intelligence tests.

We plan to answer the following research questions quantitively after completing data collection and will report on them in our talk at AHFE and in another publication (results will also be available on our OSF repository: https://osf.io/xp6c5).

H1. Reverse engineers are on average slower and perform worse on manually obfuscated code than on clear code.
H2. The higher participants' expertise, the higher their performance in obfuscating or reversing code. Further, the higher a reverse engineer's experience, the less time they require when reverse engineering code.
H3. Divergent thinking, convergent thinking, ambiguity tolerance, and the control of divergent and convergent thinking are positively related to performance when obfuscating or reversing code. Further, these variables are negatively related to time required to reverse engineer code.
H4. Participant pairs who receive a game theoretic course on adversarial reasoning outperform participants in the control condition when obfuscating or reverse engineering code.

The focus of the results section of this paper, however, shall lie on an exploratory description of the obfuscation process of selected participants given that data collection is still on-going.

4 Results and Discussion

To illustrate this results and discussion section, we have uploaded the screen capture of selected participants of our study onto our OSF repository (https://osf.io/xp6c5).

Strategies chosen by code obfuscators in our study included:

- reverse engineering and modifying the code themselves to anticipate their adversaries' behavior,
- replacing switch-loops with if-else-constructions to reduce legibility,
- changing integers (e.g. in if-queries) to mathematical operations resulting in the same value,

- randomly inserting mathematical functions to transform numbers that do not affect program logic,
- replacing variables with calls to methods which produce those same variables or values,
- adding pointless lines of code in which unused get transformed,
- changing variable names to nonsensical strings or to meaningful variables pointing at false identities of these variables,
- adding redundant boolean checks (e.g. "if not not not not not not not false, then",
- using short-ifs to place regular if-checks and the consequential behaviour in both cases into one line, and
- generally omitting line breaks.

Obfuscators often wondered what attackers might want to achieve and how they could sabotage them, i.e. they performed adversarial reasoning. Even though some have learned about level-k reasoning [11] in the course, their reasoning usually remained on level 1. They thus did not wonder what the attacker would wonder what they would do, or even wonder what the attacker would wonder what they would wonder what the attacker would wonder what they would wonder what the attacker would do. Usually, at least level-3 reasoning is necessary to be one step ahead of an adversary [8], therefore attackers often solved all reverse engineering problems despite the obfuscation. Still, in most cases this far, attackers were much slower in doing so when working on problems on obfuscated code, indicating that reverse engineering obfuscated code is indeed a more difficult task than reverse engineering clear code, even if the obfuscation has been performed manually by relatively unexperienced actors.

When observing participants in the group that did receive a game theory training on adversarial reasoning, it was even more evident that participants tried to apply the course contents by considering the adversary and their potential goals, attributes, strategies, and thoughts. One participant, for example, finished with all reverse engineering tasks after working on them for approximately 40 min, but performed several further checks for another 20 min to ensure the correctness of his solutions, fearing that the obfuscator had made it appear easy so that he would think himself safe, and being determined not to fall into that trap (note that participants are instructed that the faster they conclude participation, the better, and that they should notify the experimenter as soon as they find satisfactory answers for all tasks). Another asked the experimenter whether the attacker would be a student, and if so, of which field, as he planned to perform RSA encryption on parts of the code and feared that this would not be as effective if the attacker, like him, were a math student. Even though evidence that these types of questions occurred more often for participants who had attended the course remains anecdotal as we have only tested only 22 participants so far, a confirmation of this trend when regarding all participants would indicate that the effect of adversarial reasoning training on success on cybersecurity tasks is a fruitful field of further investigation.

Generally, a significant predictor of success was whether or not obfuscators tested their transformations regularly, e.g. by running the program to check for errors and exceptions or by printing out variables after changing code portions in order to check whether their values remain unchanged. Oftentimes, participants made several changes to then notice that something went wrong and could not retrace their steps to undo their mistakes. One can thus say that obfuscation is a delicate process in which one must constantly be on track regarding the effect one's changes have on the code. Other ways to keep up with one's own process were to note variables' "real" identities into comments after changing their names into senseless or misleading names, copying the original code into a safe file, and using strings that appear meaningless to a naïve reader, but include enough hints for the obfuscator to remember what this variable's name originally was.

We hope that this study demonstrates the value of studying code obfuscation and reverse engineering through a psychological lens. Ultimately, we believe conceptualizing these two technological procedures as problem solving processes will improve both security measures as well as development of psychological theory on cognitive abilities and applied reasoning and look forward to presenting quantitative results of our study at the AHFE 2020.

References

1. Fyrbiak, M., Strauß, S., Kison, C., Wallat, S., Elson, M., Rummel, N., Paar, C.: Hardware reverse engineering: overview and open challenges. In: 2017 IEEE 2nd International Verification and Security Workshop (IVSW), pp. 88–94. IEEE (July 2017)
2. Dörner, D., Funke, J.: Complex problem solving: what it is and what it is not. Front. Psychol. **8**, 1153 (2017)
3. Hendrix, D., Cross, J.H., Maghsoodloo, S.: The effectiveness of control structure diagrams in source code comprehension activities. IEEE Trans. Softw. Eng. **28**(5), 463–477 (2002)
4. Schrittwieser, S., Katzenbeisser, S., Kieseberg, P., Huber, M., Leithner, M., Mulazzani, M., Weippl, E.: Covert computation: hiding code in code for obfuscation purposes. In: Proceedings of the 8th ACM SIGSAC Symposium on Information, Computer and Communications Security, pp. 529–534. ACM (May 2013)
5. Park, J., Kim, H., Jeong, Y., Cho, S.J., Han, S., Park, M.: Effects of code obfuscation on android app similarity analysis. J. Wirel. Mob. Netw. Ubiquitous Comput. Dependable Appl. **6**(4), 86–98 (2015)
6. Ceccato, M., Di Penta, M., Falcarin, P., Ricca, F., Torchiano, M., Tonella, P.: A family of experiments to assess the effectiveness and efficiency of source code obfuscation techniques. Empir. Softw. Eng. **19**(4), 1040–1074 (2014)
7. Hänsch, N., Schankin, A., Protsenko, M., Freiling, F., Benenson, Z.: Programming experience might not help in comprehending obfuscated source code efficiently. In: Fourteenth Symposium on Usable Privacy and Security (SOUPS 2018), pp. 341–356 (2018)
8. Hamman, S.T., Hopkinson, K.M., Markham, R.L., Chaplik, A.M., Metzler, G.E.: Teaching game theory to improve adversarial thinking in cybersecurity students. IEEE Trans. Educ. **60**(3), 205–211 (2017)
9. Dörner, D., Kreuzig, H.W., Reither, F., Stäudel, T.: Lohhausen: Vom Umgang mit Unbestimmtheit und Komplexität. [Lohhausen: dealing with indefiniteness and complexity] (1983)
10. Summers, T.C.: How hackers think: a mixed method study of mental models and cognitive patterns of high-tech wizards (Doctoral dissertation, Case Western Reserve University) (2015)
11. Stahl, D., Wilson, P.: On players' models of other players: theory and experimental evidence. Games Econ. Behav. **10**, 218–254 (1995)

Cyberbullying Perceptions and Experiences in Diverse Youth

April Edwards[1(✉)], Lynne Edwards[2], and Alec Martin[2]

[1] Cyber Science Department, United States Naval Academy, Annapolis, MD, USA
aedwards@usna.edu
[2] Media and Communication Studies Department, Ursinus College, Collegeville, PA, USA
ledwards@ursinus.edu, alecmartin34@gmail.com

Abstract. We report results from a study that fills gaps in cyberbullying research regarding: 1) diverse youths' cyberbullying perceptions and experiences, and 2) youths' cyberbullying-related self-disclosure online. Our study surveyed the relationship between youths' online self-disclosure practices and their cyberbullying experiences, finding a significant correlation between cyberbullying and the amount and type of information disclosed online. We also found racial differences in youths' self-disclosure. Focus group discussions with LGBTQ+ youths explored these issues in more depth. We found distinctive perspectives about accountability, disclosure management, and witness-victim effects.

Keywords: Human factors · Cyberbullying · Cyber crime · User studies

1 Introduction

Recent reports by the Cyberbullying Research Center show that 17% of youths reported being cyberbullied in 2019[1]; however, this statistic hides an uglier story for some youths. LGBTQ youths report being cyberbullied at higher rates—27.1%, according to the CDC.[2] Previous research rarely shows the diversity of youths who are impacted by cyberbullying, the diversity of youth perspectives on cyberbullying, or even authentic youth voices engaged in cyberbullying or cyberbullying-related practices. In this article, we describe the results from a study on cyberbullying behavior and diverse youths. The motivation for this project was two-fold. First, we wanted to talk to youth directly about cyberbullying. Second, we wanted to be able to learn more about the attitudes, experiences and behavior among youth from different subpopulations. To that end, we obtained participation across racial/ethnic identities in an online survey and conducted a focus group with adolescent members of the LGBTQ+ community.

---

[1] https://cyberbullying.org/2019-cyberbullying-data.
[2] https://www.stopbullying.gov/bullying/lgbtq.

I. Corradini et al. (Eds.): AHFE 2020, AISC 1219, pp. 9–16, 2020.
https://doi.org/10.1007/978-3-030-52581-1_2

2 Background and Related Work

Cyberbullying is the use of social media, email, cell phones, text messages, and Internet sites to threaten, harass, embarrass, or socially exclude someone [1, 2]; its roots lie in traditional forms of relational bullying and its inherent power imbalance. The anonymity and audience-size afforded by social media and the Internet contribute to the power imbalance between cyberbullies and their victims. Cyberbullies can remain anonymous while attacking their victims and they are able to post these messages to a wide audience – much larger than the schoolyard [3–5].

Youths regularly disclose personal information to friends and strangers in their online social networks [6, 7]. These young people, particularly LGBTQ youths, find their online connections beneficial, improving their relationships with friends and reducing feelings of loneliness [8, 9]. Unfortunately, privacy research has shown that age, gender, and relationship status are related to the disclosure of highly personal information, including sexual activity among youths [10, 11].

There is a lack of research exploring the diversity of youths involved in cyberbullying [12, 13]. Lenhart, et al. [14] conducted one of the few studies that directly asked youths about malicious online behavior and disaggregated the findings by race, reporting notable racial disparities, with 72% of white teens and only 56% of black teens indicating that the way peers treat one another on social media online is "mostly kind." In a later study, more black teens (17% vs. 11% of the white youths and 4% of Latinos) reported that they "more frequently" see cruelty online [15]. Olsen et al. found that LGBTQ youths were at increased risk for bullying victimization [16].

3 Self-disclosure and Cyberbullying Behavior

In our prior work, we found that youths of color report lower levels of cyber-victimization than their white peers; indeed, black and Hispanic youth are more likely to be cyberbullies and offline bullies than victims of bullying [17]. In this section, we describe the results from an online survey that examines the relationship between self-disclosure and cyberbullying for youths of color when compared to white youth.

Our online survey included questions about cyberbullying experiences and perceptions, adapted from Willard's cyberbullying survey [1] and questions about self-disclosure attitudes and habits were adapted from Jourard's self-disclosure survey [18, 19]. We also collected standard demographic information about race, gender, and age. Face-book advertising was used to direct youth to the survey, and respondents were entered into a drawing for a gift card to thank them for their participation. Participant consent was required to enter the survey system, and additional parent/guardian consent was required for minors.

There were 221 total responses. Seventy-three surveys were omitted because they were incomplete, leaving a total of 148 completed surveys for this analysis. Participants ranged in age from 10 to 19 and older, with a median age of 15 years old. The respondents were predominantly female (64.9%), from a variety of racial and ethnic backgrounds (Table 1).

Table 1. Cyberbullying perceptions

Race/ethnicity	Reporting being cyberbullied	Reporting cyberbullying others	Reporting that friends have been cyberbullied	Reporting that friends cyberbullied others
White/Caucasian (n = 64)	48.4%	12.5%	70.3%	42.2%
African American (n = 25)	48.0%	12.0%	60.0%	52.0%
Hispanic (n = 31)	32.3%	16.1%	51.6%	35.5%
Two or more[a] (n = 18)	55.6%	22.2%	72.2%	44.4%
Other (n = 8)	75.0%	0.0%	100.0%	62.5%
Asian (n = 2)	50.0%	100.0%	50.0%	100.0%
All non-white (n = 84)	*46.4%*	*16.7%*	*63.1%*	*46.4%*
All (n = 148)	**47.3%**	**14.9%**	**66.2%**	**44.6%**

[a]Participants could select multiple identities. 14 of the 18 who selected two or more included White/Caucasian in addition to at least one other identity.

Cyberbullying Perceptions and Experiences. The survey collected information about cyberbullying experience. The results appear in Table 1. These findings suggest that Black and Hispanic youths experience cyberbullying at rates similar to those of their White counterparts, but differently from each other.

Self-disclosure Practices. Self-disclosure habits were collected by adapting Jourard's original 60-item self-disclosure survey [18], updating the language and deleting items that were not relevant to youth (e.g., items about spouses, taxes, etc.). The revised survey contained 30 items measuring respondent's willingness to disclose feelings about their appearance, personality, and interests to parents, friends, romantic partners, people online, or no one at all. Disclosure scores[3] from our respondents ranged from 0 to 104, and were then summarized by race, gender, and cyberbullying history.

As seen in Table 2, youth of color disclosed significantly less than white youth online. This is interesting when you consider that according to a 2015 PEW Research Center report, 34% of black teens and 32% of Hispanic teens are online almost constantly; while only 19% of white teens report this much usage [15, 20]. Furthermore, while all youth disclosed more to parents and friends than to romantic partners or online friends, white youth disclosed more across all types of relationships, when compared to

[3] To calculate the disclosure score, we assigned 1 point for each sharing option identified, and 0 points if participant did not report sharing with one of these target types. When respondents selected the "none" option, the entire item was scored as 0. Thus, each of the 30 items received a score from 0 to 4. Participants' responses for each item were then summed to create a scale that ranged from 0 (non-discloser) to 120 (high-discloser).

youth of color. Unsurprisingly, Table 2 shows that all youth are more likely to disclose information about Tastes and Interests than more personal information, such as concerns about personality or thoughts about their bodies. Youth of color again differed from their white peers across all categories. Statistical testing was done to determine if there were racial differences related to disclosure, and to determine the relationship between being cyberbullied and disclosure. We used a two sample two-tailed t-test to determine level of significance across categories. White youth disclosed significant more when compared to youth of color ($p < .05$). Those who had been cyberbullied reported significantly higher disclosure rates than those who had not ($p < .005$), as shown in Table 2.

Discussion. Our findings confirm that youths who report lower self-disclosure rates also report fewer cyberbullying experiences. There are clear racial/ethnic differences in self-disclosure behaviors, with White youths disclosing more types of personal information to more people, online and offline, than non-White youths. Given that youth of color report lower levels of victimization among both themselves and their friends (Table 1), they may be limiting disclosure as a preventative measure.

Table 2. Disclosure score findings (min = 0, max = 120), all values are average disclosure score

Race/ Ethnicity	Parent	Friend	Romantic Partner	Online Friend	No One	Total
Non-white	11.9	15.2	8.1	5.0	8.9	39.0*
White	14.8	17.6	9.8	5.8	7.0	46.7*
All	**13.1**	**16.2**	**8.8**	**5.3**	**8.1**	**42.3**

Race/Ethnicity	Tastes and Interest	Personality	Body Perceptions	Total
Non-white	17.7	12.3	9.0	39.0*
White	20.5	13.8	12.4	46.7*
All	**18.9**	**13.0**	**10.5**	**42.3**

Cyberbullying History	Average Disclosure Total
Was Cyberbullied	47.9+
Was not Cyberbulllied	35.1+
Not sure	44.3
All	**42.3**

*difference is significant p<.05
+difference is significant p<.005

4 Focus Group with LGBTQ+ Youths

We partnered with a local counseling center that offers support groups for children and youth to recruit participants for our LGBTQ+ focus group. The participants were informed of the risks of the study and provided informed consent to participate. When participants were under the age of 18, parents or guardians were also required to provide consent for participation. The participants were offered a $25.00 gift card to thank them for their time, and pizza and soda were served during the focus group session. Participants were known to each other and comfortable sharing personal information, due to their prior experience at the counseling center.

We trained two undergraduate research assistants to facilitate the focus group. Not only did this provide good research experience for the students, we believe that having facilitators closer in age to the participants helped reduce barriers. As an extra precaution, one staff member from the counseling center remained unobtrusively in the room in case the discussion triggered a negative event in any of the participants. The principal investigators, parents and other center staff remained nearby but were excluded from the room during the sessions.

Focus group participants were led through a guided script with four multi-part questions that asked participants to define cyberbullying, if they perceive cyberbullying as a problem, what they think of when someone says self-disclosure and if there is a connection between disclosing information online and cyberbullying. The session lasted about 60 min and was recorded on two digital recording devices. The participants were aware of, and consented to, the recording. The recordings were later transcribed and coded for content and tone.

The six participants in the focus group were between the ages of 15 and 19, inclusive and identified as members of the LGBTQ+ community. When asked to define cyberbullying and their impressions of cyberbullying, participants were remarkedly consistent with the literature, using language similar to "*[i]t's bullying either over like texting, or like harassments or threats via the Internet, or over mobile devices.*" After analysis of the transcripts, we identified two critical themes, not previously mentioned in the literature, surrounding cyberbullying and self-disclosure: activity abandonment, and active self-disclosure management.

Activity Abandonment. One result of cyberbullying was activity abandonment or hesitation to begin an activity all together. One participant gave a detailed account of why he decided against sharing his activity on YouTube: "*originally I wanted to do some stuff on YouTube ... but I have seen something that makes me hesitant about actually doing it. I'm an equestrian ... and there's a lot of people who will search out videos just to shame you on your equitation.*" It is interesting that just witnessing cyberbullying behavior was enough to change the behavior of this participant. Another participant was an active user of a fan fiction writing community, and claimed that he stopped writing for two years because of another user who "*literally went through, on every single one of my stories and posted some sort of really rude comment about how I was a shitty writer, and how I should go kill myself.*" For these participants, both the potential for cyberbullying and prior experience with cyberbullying were reasons to modify their online behavior - censuring themselves to reduce their exposure to cyberbullying.

Active Disclosure Management. None of the participants said that they were the target of homophobic or sexual orientation-based cyberbullying online, perhaps because they went to great lengths to conceal their personal information in their online identities. One participant shared her perspective: "*... the more information they have on you the more ammunition they have*" Another participant added: "*defining yourself sexually can* *like give so much ammunition to people that are really out there to hurt you*" The use of the word '*ammunition*' is particularly troubling, suggesting a figurative death-threat for their authentic identities. When the moderator phrased a scenario about someone telling a friend in person that he was gay, the participants agreed that sharing this would be okay if the two are close friends. The group also agreed that someone should not share anything online that they did not want in public and went on to describe savvy ways to conceal their identities online. This discussion helped shed light as to why these participants did not mention being personally cyberbullied because of their sexual identity; they are going to extreme lengths just to feel secure with their friends on the Internet. One participant elaborated "*it depends on who you're talking to like say there's this person you Skype with every single day … It's someone you can probably generally trust to know some personal information, probably not all of it.*"

For youths struggling with issues of gender and sexuality, one would presume that personal information, in any setting, would have to do with their gender and sexual identity, but it is clear that these participants, regardless of whether they stated they had a personal experience with cyberbullying, shy away from revealing any type of personal information online, including name, location, and place of work.

Discussion. Previous research found that sexual minorities were the largest target for bullying [16, 21]. Although the perceptions of cyberbullying and its effects are consistent with previous, survey-based research, the findings related to the personal impact of cyberbullying on our participants who witnessed cyberbullying are novel. One result of witnessing cyberbullying was activity abandonment or hesitation by study participants, a result not observed in previous quantitative studies. The personal accounts of activity abandonment, as well as the limited data on homophobic and sexual orientation-based cyberbullying help show how these adolescents are actively managing their self-disclosure online.

5 Conclusions and Future Work

We have reported our findings from a study to investigate the online experiences of diverse youth, specifically experiences related to cyberbullying behavior and online disclosure of personal information. Using an online survey, we found a significant relationship between online self-disclosure and cyberbullying experience, as well as differences in self-disclosure behavior between white youth and youth of color. We also explored attitudes and behaviors around cyberbullying with an LGBTQ focus group and found evidence of activity abandonment and self-disclosure management from witnessing online bullying. We found pervasive attitudes that these youths need to actively manage what they disclose online to avoid becoming victims. as well as the possibility of negative effects from merely witnessing cyberbullying.

Acknowledgments. We thank our participants for their insights and participation in this important work. This material is based upon work supported by the National Science Foundation under Grant Nos. 0916152 and 1421896. Any opinions, findings, and conclusions or recommendations expressed in this material are those of the author(s) and do not necessarily reflect the views of the National Science Foundation.

References

1. Willard, N.: Cyberbullying and Cyberthreats: Responding to the Challenge of Online Social Aggression, Threats, and Distress. Research Press, Champaign (2007)
2. Olweus, D.: Bullying at School: What We Know and What We Can Do. Blackwell, Oxford (1993)
3. Dempsey, A.G., Sulkowski, M.L., Nichols, R., Storch, E.A.: Differences between peer victimization in cyber and physical settings and associated psychosocial adjustment in early adolescence. Psychol. Sch. **46**, 962–972 (2009)
4. Schneider, S.K., O'Donnell, L., Stueve, A., Coulter, R.W.: Cyberbullying, school bullying, and psychological distress: a regional census of high school students. Am. J. Public Health **102**, 171–283 (2012)
5. Horowitz, M., Bollinger, D.M.: Cyberbullying in Social Media within Educational Institutions. Roman & Littlefield, New York (2014)
6. Youn, S.: Teenagers' perceptions of online privacy and coping behaviors: a risk-benefit appraisal approach. J. Broadcast. Electron. Media **49**(1), 86–110 (2005)
7. Mishna, F., McLuckie, A., Saini, M.: Real world dangers in an online reality: a qualitative study examining online relationships and cyber abuse. Soc. Work Res. **33**(2), 107–118 (2009)
8. Lenhart, A., Madden, M.: Social networking websites and teens: an overview. Pew Internet & American Life Project Teens and Parents Survey, October-November 2006. http://www.pewinternet.org/PPF/r/211/report_display. Accessed 7 Jan 2007
9. Ellison, N., Steinfield, C., Lampe, C.: Connection strategies: social capital implications of Facebook-enabled communication practices. New Media & Society (2011). https://doi.org/10.1177/1461444810385389. Accessed 27 Jan 2011
10. Nosko, A., Wood, E., Molema, S.: All about me: disclosure in online social networking profiles: the case of Facebook. Comput. Hum. Behav. **26**(3), 406–418 (2010). https://doi.org/10.1016/j.chb.2009.11.012. ISSN 0747-5632
11. Wan, C., Chung, S., Chiou, W.: Contingent impression management in sexual disclosure by older adolescents corresponding in cyberspace: the role of gender dyads. Soc. Behav. Personal. **37**(8), 1023–1032 (2009)
12. Carter, J.M., Wilson, F.L.: Cyberbullying: a 21st century health care phenomenon. Pediatr. Nurs. **41**(3), 115–125 (2015)
13. Low, S., Espelage, D.: Differentiating cyber bullying perpetration from non-physical bullying: commonalities across race, individual, and family predictors. Psychol. Violence **3**(1), 39–52 (2013). https://doi.org/10.1037/a0030308
14. Lenhart, A., Madden, M., Smith, A., Purcell, K., Zickuhr, K., Rainie, L.: Teens, kindness and cruelty on social network sites. Pew Research Center's Internet and American Life Project (2011). http://www.pewinternet.org/2011/11/09/teens-kindness-and-cruelty-on-social-network-sites/
15. Lenhart, A., Smith, A., Anderson, M., Duggan, M., Perrin, A.: Teens, technology, and friendships. Pew Research Center (2015). http://www.pe winternet.org/files/2015/08/Teens-and-Friendships-301 FINAL2.pdf
16. Olsen, E.O., Kann, L., Vivolo-Kantor, A., Kinchen, S., McManus, T.: School violence and bullying among sexual minority high school students, 2009–2011. J. Adolesc. Health **55**(3), 1–7 (2014)
17. Edwards, L., Kontostathis, A., Fisher, C.: Race, cyberbullying and mental health: a review of the literature. Media Commun. **4**(3), 71–78 (2016)
18. Jourard, S.M.: Self-Disclosure: An Experimental Analysis of the Transparent Self. Wiley-Interscience, New York (1971)

19. Jourard, S.M., Lasakow, P.: Some factors in self-disclosure. J. Abnorm. Soc. Psychol. **56**, 91–98 (1958)
20. Tynes, B.M., Mitchell, K.J.: Black youth beyond the digital divide: age and gender differences in Internet use, communication patterns, and victimization experiences. J. Black Psychol. **40**(3), 291–307 (2014). https://doi.org/10.1177/0095798413487555
21. Mueller, A., James, W., Abrutyn, S., Levin, M.L.: Suicide ideation and bullying among us adolescents: examining the intersections of sexual orientation, gender, and race/ethnicity. Am. J. Public Health **105**(5), 980–985 (2015)

Ethics, Economics, and Ransomware: How Human Decisions Grow the Threat

John Christian Bambenek(✉) and Masooda Bashir

University of Illinois at Urbana-Champaign, Champaign, IL, USA
{bambenek,mnb}@illinois.edu

Abstract. This paper examines the modern history of ransomware and its evolution to the current form of large-scale ransomware attacks (ones that disrupt entire organizations). Within that timeframe, public reporting, articles, and news media reporting on large-scale ransomware attacks is reviewed to create an empirical analysis of ransom payments, conditions that led to those payments, and if data was ultimately recovered.

Three factors were discovered that lead to organization to pay the ransom when recovery is impossible or cost-prohibitive: the rise of cyberinsurance companies that dictate responses that lessen their financial exposure, many victim organizations who have to always operate such as hospitals and emergency services, and the fiduciary duty of business executives to act in the best interest of a company. Lastly, we look at the concept of outlawing ransom payments and relate it the policy of outlawing random payments in kidnapping.

Keywords: Human factors · Ransomware · Cybersecurity · Incident response

1 Introduction

In 2013, the first successful mainstream ransomware called Cryptolocker spread across the Internet. Since then, the threat has grown and is now a common-place incident making headlines routinely. Among the concerns that are routinely expressed is the ethical considerations of paying ransoms and how those who do pay are merely funding the next attacks. On one hand, limited the profitability of such attacks would lessen their occurrence. On the other hand, it would require organizations to accept the permanent loss of data and potentially shut down entirely.

This paper provides an empirical study of 206 public reports from 2018 and 2019 involving over 1,100 affected organizations to identify payments by those organizations and attempt to identify the context and circumstances that led to those payments. Also, in the light of recently introduced legislation to prohibit ransomware payments, this policy is examined alongside the policy of the nation of Colombia in banning ransom payments for kidnapping and the factors that made the latter policy successful.

What was found is that the human element involved in decision-making when deciding to pay a ransom almost always highly-incentivizes the decision to pay as long as

I. Corradini et al. (Eds.): AHFE 2020, AISC 1219, pp. 17–22, 2020.
https://doi.org/10.1007/978-3-030-52581-1_3

there is confidence of a recovery. Between cyberinsurance companies wanting to lower costs, executives having a fiduciary duty to limit the cost of an incident, and that a subset of victims provide emergency services that always must be available, human nature dictates almost always that ransoms will be paid.

2 Evolution of Ransomware

The first known incident of ransomware was the AIDS Trojan distributed via floppy disks in 1989. It required victims to mail the ransom to a post office box in Panama [1]. It wasn't until 2013 with the emergence of Cryptolocker that ransomware took off. By using bitcoin as a payment mechanism and using modern encryption and solid encryption key management, attackers found the perfect mix of technologies to make ransomware at scale possible.

The earlier forms of ransomware infected single machines and typically demanded ransoms between $300 and $1000 in bitcoin. These were often spread by e-mail or web-based exploit kits and abused various flaws in operating systems [2].

The WannaCry and NotPetya ransomware attacks highlighted an emerging form of ransomware, namely, ransomware that doesn't infect single machines, but attempts to debilitate an entire business. Ryuk, SamSam, and Sodinokibi families of ransomware are more recent malware families designed to cripple entire organizations and use a variety of means to spread [3]. This means attacks can demand higher ransoms as the stakes are much higher.

3 Methodology

This paper examined 206 public reports dating from January 1, 2018 onward. Many of these reports were summarized in a Ransomware Attack Map hosted by Cyberscoop [4]. These reports covered over 1,100 affected organizations as some reports covered multiple incidences of the same attack. These were found by looking for ransomware in news media reporting and, where available, examining underlying documents and reports. Of those, only ransomware incidents that affected an entire organization were examined as opposed to single machines that were affected.

Of those reports, the number of organizations was identified, the industry vertical the organization represents, whether they paid a ransom or not, and if there were reasons given for paying that ransom that might shine a light into the human factors involved in that decision-making. In most media reporting of the decision to pay the ransom or not, an official has made a comment on why they made the decision they did.

It is important to note that this analysis is imperfect as organizations generally are not obligated to report nor are they compelled to completely disclose the nature of the incident or its details. Many public reports leave out elements that would be useful to this analysis such as whether they paid the ransom or not, or even the size of the ransom demand.

4 Results

Of the reports examined, approximately 78% of publicly confirmed ransomware cases there was no reporting on whether a ransom was paid or not. Of the remaining incidents, 14.5% did not pay the ransom and 7.5% did pay the ransom. It should be noted that there are many unreported cases of ransomware that occurred that cannot be effectively studied.

A large body of public incidents that are reported on involve government as local government activity has to be largely public with 173 reports. Healthcare also was a large contributor of cases with 766, likely because disruptive impact to a healthcare institution is readily noticed and regulatory requirements mandate disclosure. Educational institutions made up 89 cases, likewise because of their public nature disruption would be obvious. In the given time period, only 2 financial institutions publicly reported ransomware. This would indicate that absent a regulatory reason or that a ransomware disruption would be obvious to the public anyway, that many institutions do not acknowledge ransomware infections.

In the public reports where the size of the ransom was mentioned, there was 1 ransom that was less than $1,000, 8 between $1,000–$9,999, 16 between $10,000 and $99,999, 8 between $100,000 and $999,999, and 12 that were larger than $1 million. The smaller the ransoms where, the more likely payment was to occur. The highest ransom payment that was recorded was $594,000.

In cases where the ransom was paid, it was mentioned most frequently that the cost of recovery was higher than the cost of the ransom. Specifically, case of the ransomware infection of Atlanta was mentioned where the ransom payment was $52,000 but the cost spent by the city to recover was $9.5 million [5]. Also mentioned was a desire to return to normal operations, particularly in health care and government, where services must always be available. In two cases, insurance companies were mentioned who desired to keep their exposure to claims as low as possible.

In a similar case, the City of Baltimore had a ransomware infection and faced a demand of $76,000. The opted not to pay the ransom and public reports indicated the cost of recovery was approximately $10 million with an additional $8 million in lost or deferred revenue due to inability to process payments. These two cases figured highly in discussions of other government ransomware incidents, but also impacted how corporations viewed the incidents.

Of those who elected to not pay the ransom, it was mentioned that it was due to law enforcement advising the victim not to pay and the desire not to contribute to the incentives of future ransomware attacks. While it is not known what the victims who acknowledged they had a ransomware attack but did not state if they paid the ransom or not eventually did, it seems likely that many of them paid.

What is notable about these results is even where public reports were available, there were many questions left unanswered. Notably the question of whether the ransom was paid or not was conspicuously absent when organizations self-reported a ransomware incident. Governments, of course, have little to no ability to make secret payments, but corporations have little duty to disclose such payments except in the notable case of publicly-traded companies. In those cases, their reports often will detail the costs of not paying ransoms as they did with WannaCry and NotPetya, but those financial reports often would not require disclosure of the relatively modest ransom payments discussed here.

The fact that few organizations publicly report ransomware incidents suggests the problem is far worse, but victims are loathe to talk about them openly for fear of reputational damage or potentially retaliation by the criminals themselves.

4.1 Impacts of Ransomware Infection

Largely, ransomware is viewed from the prism of financial circumstances. The ransom is monetary, and the costs involved with recovery are monetary. It is noteworthy to mention, there are four cases in which ransomware-related attacks lead to the closure of a corporation. The first was a company called Code Spaces. A hacker gained access to their infrastructure and demanded a ransom. When it wasn't paid, they deleted all their storage and customer information which lead them to immediately close [6].

In the second incident, a fundraising firm The Heritage Company suspended operations and laid off their entire staff even though they had paid the ransom. The third and fourth incident involved two small medical offices that opted to close after declining to pay the ransom. In a subset of cases, a ransom payment may mean the difference between a business continuing to exist or to close.

There are also other non-financial interests to consider. A recent study into the effects of ransomware attacks on hospitals showed that hospitals that suffered breaches including ransomware often had longer times to providing critical services (such as EKGs) that has led to a measurable increase in mortality rates of those facilities compared to those that did not suffer a breach or ransomware infection [7].

Government institutions were more likely to not pay the ransom as closure is not a plausible outcome for them. They are also not bound by fiduciary duty the way corporate offices and boards are which would dictate that act in the best interests of shareholders even at the cost of a policy decision that may help broader society (such as minimizing the incentive for ransomware operators by not paying ransoms). Additionally, they have means of raising money not available to corporations in the form of raising taxes.

As an example, the City of Atlanta had a ransom demand of $52,000 compared to a recovery cost of $9.5 million. It is highly unlikely anyone would choose to pay two orders of magnitude more money assuming that the ransom payment can reasonably be assured to result in recovery.

There is also the personal impact of executives in charge of IT or IT security on their future careers and the intangible costs to organizations for reputational damage that may occurs as being identified as a victim of ransomware. While impossible to quantify, those making decisions in addressing ransomware have the obvious tangible costs mentioned above, they are likely to engage in decisions that do not adversely affect their employability or company's future reputation unnecessarily. It is reasonable to assume the reason that most public reports for ransomware infections involve certain industries is that many have no obligation to report and have the ability to hide the disruption from the public, thus they have no incentive to disclose.

5 Outlawing Ransom Payments

Recently public policy discussion and ethical discussion regarding ransomware suggests that some believe victims should adopt these higher costs of not paying the ransom and paying recovery costs because ransom payments incentivize the attackers to keep attacking.

As a useful analogy, ransoms in kidnapping may be worth examining. For the most part, there are few countries with general laws prohibiting the payment of ransoms in kidnapping. One notable exception is the nation of Columbia. As part of its effort to tackle kidnapping, there was a law prohibiting payment of ransoms that was eventually declared unconstitutional. Additionally, there were efforts to retake territory controlled by guerilla forces and setting up specialized law enforcement units to handle kidnapping. This led to a dramatic reduction in kidnapping in a country that once was known as the kidnapping capital of the world [8].

In the general case, kidnapping is more severe than ransomware, but in extreme cases, the impact of large-scale ransomware could lead to far larger losses of life. While there isn't a large body of research on kidnapping policies generally, in this particular case, Columbia's reduction in kidnapping shouldn't be attributed to the prohibition on paying ransoms. What is more likely is that the large amount of other government actions to tackle the problem is what ultimately led to its reduction there.

A significant difference between kidnapping and ransomware is that measures to prevent kidnapping are known and able to be taken by a government. A victim and perpetrator are in the same location. This means a variety of physical security measures and traditional law enforcement tools are able to be undertaken assuming the government is in a position to do so.

On the other hand, ransomware can flourish because the consequences of cybercrime are often difficult if not impossible to obtain. A perpetrator can operate in a different part of the world than the victim and the presence of tools to move money (cryptocurrency) outside the banking system in an anonymous fashion means traditional law enforcement tools are more difficult to employ. International law enforcement is, at best, difficult and there are countries who essentially do not cooperate on cybercrime matters.

It is also those same anonymous payment methods which allow victims to nonetheless pay ransoms and if their breach was never known, there would be no way to enforce a "no ransom payments" policy. Even if it were possible to know if a ransom payment were made, unless the penalty for payment were consummate with the impact of not paying the ransom, it may be financially in the best interest of an organization to pay the ransom and deal with the consequences.

For instance, in the City of Atlanta case, if decision-makers knew in advance recovery would cost $9 million and the ransom payment was $52,000, they could have elected to pay the ransom anyway. If the penalty was less than $9 million, it may make financial sense to accept a fine. If the penalties for paying ransom were some criminal sanction, and the case could involve a loss of life, individuals may decide that the value of saving lives may be more important than the consequences of incarceration, assuming such laws could even be passed and withstand court scrutiny.

6 Conclusion

Based on public reports, it's possible to conclude human and economic decisions are such that the advantage is in the hands of ransomware operators. For organizations with a fiduciary duty, as long as the cost of the ransom is lower (especially much lower) than the cost of recovery and the likelihood of recovery after paying the ransomware is high, it makes financial sense to pay the ransoms. When loss of life is a possibility, human nature will highly incentivize those concerns over financial penalties. With the presence of cyberinsurance companies having to foot the bill, they will opt for lower payments in ransoms than larger claims in recovery costs.

The notable deviation is government agencies who don't face existential risks due to ransomware. Local governments will not cease to exist, and they are not subject to market forces. If a government needs more money, they can increase taxes or sell bonds; these options are not available to corporations.

Lastly, in examining kidnapping and ransoms in Colombia, the prohibition of ransom payments did not have a large impact in reducing kidnapping by eliminating the financial incentive. What did work was a large investment by the government to tackle the ability of groups to conduct kidnapping operations. Unfortunately, analogous techniques with ransomware are not readily available at this time. What is clear is that reducing incentives of this behavior likely will not work unless they are targeted at increasing consequences to the ransomware operators as opposed to their victims.

References

1. Duggan, M.: The legal corner (TLC): ransomware attacks against health care IT. J. Inform. Nurs. **2**(4), 30–31 (2017)
2. Goel, S., Bambenek, J., Bashir, M.: Ransomware: recommendations against the extortion. In: 11th IADIS International Computer Information Systems 2018, pp. 193–200. IADIS (2018)
3. Hassan, N.: Ransomware Revealed, pp. 47–68. Apress, Berkeley (2019)
4. Cyberscoop: Ransomware Attacks Map (2020). https://statescoop.com/ransomware-map/. Accessed 1 Feb 2020
5. Kierney, L.: Atlanta officials reveal worsening effects of cyber attack. Thomson Reuters (2018). https://www.reuters.com/article/us-usa-cyber-atlanta-budget/atlanta-officials-reveal-worsening-effects-of-cyber-attack-idUSKCN1J231M. Accessed 1 Feb 2020
6. Chen, J.: Cyber security: bull's-eye on small businesses. J. Int. Bus. Law **16**, 97 (2016)
7. Choi, S.J., Johnson, M.E., Lehmann, C.U.: Data breach remediation efforts and their implications for hospital quality. Health Serv. Res. **54**(5), 971–980 (2019)
8. Gurney, K.: Behind Columbia's dramatic fall in kidnappings. InSight Crime (2015). https://www.insightcrime.org/news/analysis/behind-colombia-dramatic-fall-in-kidnappings/. Accessed 1 Feb 2020

Does the Propensity to Take Risks Influence Human Interactions with Autonomous Systems?

Priscilla Ferronato[1] and Masooda Bashir[2](✉)

[1] Illinois Informatics Institute,
University of Illinois at Urbana-Champaign, Champaign, IL, USA
pf4@illinois.edu
[2] School of Information Science,
University of Illinois at Urbana-Champaign, Champaign, IL, USA
mnb@illinois.edu

Abstract. Technological development towards automation has been taking place for years and a wide range of autonomous systems (AS) have been introduced in homes and retailing spaces. Although these AS seem to be riskless, if they are exploited they can endanger private information of users, which opens a new stage for the security of AS. Humans have an initial and positive bias towards automation that might lead to errors related to unintentional actions or lack of actions. Therefore, the effective adoption of AS relies on users' attitudes, like the propensity to take risks and the calibration of human trust to avoid situations of mistrust, over trust, and distrust, increasing the systems' security. This study conducted an online questionnaire to investigate the relationship between an individual's propensity to take risks and trust in automation. We found that participants with low risk seeking tendencies will trust more in AS when compared to high risk seeking participants. Moreover, other individual differences like age, gender, and education led to interesting results. Thus, our study provides valuable information about the human factors that mediate human and autonomous systems interactions and thereby influence trust.

Keywords: Autonomous systems · Risk · Trust in automation · Cybersecurity · Human factors

1 Introduction

Autonomous systems (AS) play an essential role in our lives performing countless tasks for which humans once had responsibility and bringing numerous benefits like when replacing humans in hazardous environments. Technological development towards automation has been taking place for years and a wide range of AS have been introduced in homes and retailing spaces. They can be used as mobile teleconference platforms, welcoming assistants, virtual pets, toys, etc. Although these AS seem to be riskless, if they are exploited, they can provide a lot of private information about users. This information

I. Corradini et al. (Eds.): AHFE 2020, AISC 1219, pp. 23–29, 2020.
https://doi.org/10.1007/978-3-030-52581-1_4

can go from general data (age, size, etc.), private pictures, user routine information, economic, etc., which opens a new set of security vulnerabilities in AS [1].

Just like in any other human type of interaction, trust is essential to achieve optimal performance in joint human-AS interactions, and an appropriate level of human trust is essential to avoid the misuse or abuse of automation. Contrarily from interpersonal trust, users' initial trust in AS is based on faith [12], and novel interactions; people often exhibit a positive bias in trusting an AS because they expect it will not fail. This initial and positive bias might lead to security related human errors - unintentional actions or lack of actions. Therefore, the effective adoption of AS relies on users' attitudes, such as the propensity to take risks [8], and the calibration of human trust to avoid situations of mistrust, over trust, and distrust increasing the systems' cybersecurity.

While human's trust in automation has been previously studied, as well as the perception of risk in AS [5], it is to the best of our knowledge that no empirical studies are examining the propensity of taking risks as an integrated part of humans' dispositional trust. Drawing from the theoretical model of trust in automation proposed by Hoff and Bashir [5], the propensity to take risks and the perception of risks in AS are two different factors. The perception of risks is an important element of situational trust, which relates to the context of the interaction and how these context-dependent variations relates to an operator's momentary mental state. Rather, the propensity to take risks relates to dispositional trust, as the individual's enduring tendency to trust in automation.

In this study, we investigate the relation between propensity to take risks [14] and trust in AS and individuals, with differences – gender, age, and education through an online survey to answer the research question: Does the propensity to take risks influence the dispositional trust in AS? The results will provide further support for the inclusion of human factors in developing AS that are human-centered, not only focusing on technology improvements such as system transparency but to avoid the misestimation and underestimation of automation which may be the root cause of cybersecurity accidents related to human and AS interaction.

2 Research Background

If trust exceeds the capabilities of the system, it will lead to over trust and the misuse of the system; if trust falls below the capabilities, it will lead to distrust, and consequently, the disuse of automation. Therefore, human operators with appropriate levels of trust in AS can reduce the frequency of misuse and disuse of automation [8]. A better understanding of users' trust can help to improve AS's security in various ways, since people differ in their ability to correctly assess risks [9].

In this research, we investigate trust in AS as a dynamic process different from interpersonal trust, since they depend on a different set of attributes. Whereas interpersonal trust can be based on the ability, integrity, or benevolence of a trustee, the human-AS trust depends on the performance, process, or purpose of an AS [5]. This adopted theoretical model is composed of three dimensions: dispositional, situational, and learned trust. Dispositional trust represents an individual's enduring tendency to trust automation. Situational trust depends on the specific context of an interaction. The environment exerts a strong influence on situational trust, but context-dependent variations in an operator's mental state can also alter situational trust. The final dimension, learned trust, is based on past experiences relevant to a specific AS, and it is closely related to Situational trust, since it is guided by past experience [13]. Although we are aware of the multidimensionality of trust in automation, in this research we concentrate our efforts to better understand an important and underrated human factor related to the dispositional dimension – the propensity to take risks.

Risk, as the degree of uncertainty associated with a given situation, is vital to understanding trust in AS [15]. Thus, an individual's perception of risks is defined as her assessment of how risky a situation is in terms of probabilistic estimates of the degree of situational uncertainty, how controllable that uncertainty is, and confidence in those estimates [18]. This is a topic regularly explored by academic and non-academic research to better understand the public trust, acceptance, and opinion about AS. As previously mentioned, the perception of risks in AS is related to the situational trust dimension, which means that the perception of risk in the system may change during the course of the interaction (e.g., unexpected system failure, weather conditions, user's level of stress). Rather, the propensity to take risks is defined as the individual current tendency to avoid or take risks [10]. Because risk propensity is cumulative, and simultaneously persistent and possible to change over time as a result of experience, it is an emergent property of a decision maker [18]. As the opposite of the perception of taking risks, this tendency does not change throughout initial interactions with automation, instead it requires time and multiple iterations. Previous research suggests that individual differences such as age, gender, and personality traits, are sources of influence on the tendency to take or avoid risks [14]. For instance, an individual risky security behavior is connected to the over-trust of automated technologies, and when the user trusts the AS too much, they will be more prone to cyberattacks [7]. The following section presents a brief review of previous studies that investigated how human factors relate to the propensity of taking risks and how it has been applied to mitigate states of over trust, mistrust, or distrust in AS.

3 Previous Findings

To the best of our knowledge, no study has investigated the influence of an individual's propensity to take risks on their dispositional trust in AS. However, others have investigated participants' perception of risk on trust in AS, which has resulted in mixed results. Perkins and colleagues [16] investigated whether trust in an automated navigation system would be affected by different levels of risk. Their results showed that participants trusted the navigation system less when more hazards were perceived. In contrast, participants in another study relied more on an automated aid compared to a human aid in high-risk conditions [11], demonstrating an inclination towards automation with the increased perception of risks. Although the perception of risk and the propensity to take risks are different attributes, we hypothesize that participants who present a higher propensity to take risks also trust more in AS given the uncertainties related to its type of new technology. Regarding individual differences, gender has been a variable widely explored, and men tend to have higher levels of risk propensity than women [3]. This is a difference that is consistent across time and in a variety of contexts, including the context of AS, in which previous studies found that men are more receptive to AS than women [4]. Therefore, the hypothesis to be tested is that men take more risks and trust more in AS as compared to women Previous investigations into the relationship between age and trust in AS show conflicting results. The majority of these studies claim that younger adults are generally more comfortable, receptive, and trust more in AS as compared to older adults [2]. Rather, Hulse and colleagues [6] noticed the oldest respondents of their survey (60+ years old) and the youngest ones (21–34) expressed the highest willingness to pay for specific AS technologies. However, another study suggested that propensity for risky behavior increases in adolescence, peaks in young adulthood, and declines with age [3]. We hypothesize that younger participants are also more willing to take risks and trust more in AS than the other participants because of its familiarity and an understanding of the autonomous technology.

Participants with higher education were associated with higher levels of trust in AS and higher acceptance rates [4]. The relationship between education and propensity to take risks has been investigated in specific fields, like management and decision making, which noticed that people with higher educational levels are more willing to take risks when associated to their area of specialization. Hence, we hypothesize that higher educational level is related to higher trust in AS and higher propensity to take risks.

4 Method

To empirically test these hypotheses and answer the research question, we conducted an online survey that consisted of four different measures. Since there is no standardized and validated measurement for dispositional trust in automation or a specific tool to measure trust in AS, we adopted the 12 items from Singh et al. [17], which is the most widely used measurement available to assess people's propensity to trust in automation. To measure the propensity to take risks, we adopted the 7 items from the Risk Propensity Scale [14]. Individual factor questions were asked to identify individual differences previously mentioned.

Four hundred participants residing in the United Stated were recruited via Amazon Mechanical Turk, and we accepted three hundred and forty-four responses ($N = 344$) based on a validation criteria. 48.58% of the respondents were female and 54.06% were between 23 and 38 years old. For education, 59.29% of participants held a bachelor's degree or higher. Participants were given $1 (US) for their participation after completing the survey. To better analyze the age groups, we classified the participants as Gen Z (up to 22 y o), Millennials (23–38y.o), GenX (39–54 y.o), Boomers (55–73 y.o), and Silent Generation (74–91 y.o), as adapted from Strauss and Howe [19].

5 Results

We ran independent linear regressions to investigate the correlation between trust in AS and propensity to take risks. In addition, we investigated if age, gender, and education are correlated with them. We found a negative correlation between trust in AS and propensity to take risks ($p = 9.07 * 10^{-06}$) and a weak correlation with age ($p = 0.0348$), in which GenX participants (39–54) are the ones with lower levels of trust. No significant correlation between gender and propensity to trust in AS were found ($p = 0.649$), and both male and female followed a similar pattern regarding their trust scores. Although no correlation was found between trust in AS and education, our dataset shows that participants with advanced degrees (master or doctorate degree) and those with less than high school are the ones with the lowest scores on trust in AS. Regarding the propensity to take risks, we found significant correlation with gender ($p = 3.46 * 10^{-05}$), in which men exhibited a higher tendency to take risks. Age also presented significant correlation ($p = 0.000143$), and Millennial and GenZ respondents were the ones more willing to take risks. No correlation was observed between risk and education (Table 1).

Table 1. Estimated parameters of linear regression models for trust in AS and propensity to take risks

Dependable variable: trust in AS					Dependable variable: propensity to take risks				
Predictors	SE	t	p-value	Adjusted R2	Predictors	SE	t	p-value	Adjusted R2
Risk	0.06999	−4.506	$9.071 * 10^{-06}$	0.05329	–	–	–	–	–
Age: GenX	1.877	−2.199	0.0348	0.01804	Age: Millenials	0.9578	3.847	0.000143	0.05806
Gender: Male	0.8608	−0.612	0.541	−0.003318	Gender: Male	0.6306	4.196	$3.46 * 10^{-05}$	0.04632
Education	1.4916	1.524	0.05716	0.02087	Education	1.13078	1.318	0.3898	0.001432

6 Discussion

This study employed self-report questionnaires to investigate if the influence of an individual's propensity to take risks influences trust in human-AS interactions, as a dispositional trust factor. We postulated that propensity to take risks and individual differences – age, gender, and education – are humans' dispositions correlated to trust in AS. Furthermore, we also hypothesized that these individual differences are correlated with participants' propensity to take risks. We confirm that propensity to take risks is statically significant and negative correlated to trust in AS, and higher the propensity to take risks, the lower is the trust in AS. This result corroborates previous findings that trust decreases with increased risk. In other words, participants with low risk seeking tendencies will trust more in AS when compared to high risk seeking participants. The tendency of taking risks does not mean that individual trust in the technology involved in the given risky situation. Therefore, further studies must be developed to better understand the differences, similarities, and the effects of propensity to take risks and the perception of risks in AS.

Regarding individual differences, we found different results from previous studies. Our results show that although males have slightly higher scores on trust in AS, they presented the same pattern of trust score distribution as females. Moreover, we found no significant correlation between gender and trust in AS. We claim that before completely rejecting the hypothesis that gender does not present a significant correlation with trust in AS, further investigation must be done with the application of different statistical models and the replication of the survey with a more diverse sample. We also could not accept the hypothesis that higher the obtained education level, the higher would be the trust in AS and the propensity to take risks. In the present study, we find no significant correlation between education level and trust in AS. In fact, we found that both participants with advanced degrees (masters and doctorate) and participants who do not finish high school presented the lowest score on trust. In contrast, participants with a bachelor's degree were the ones with the highest trust score. This is an interesting finding that must be further investigated since it can be associated to different trust components. First, a greater amount of specialized information might decrease trust in AS, as well as the lack of knowledge, in which the variable is not education but the knowledge calibration. Second, advanced degree attainment might be related to older participants, like GenX who presented the lowest level of trust in AS; thus the variable is not related to education but age. Importantly, our results show that Gen Z, male, and with a bachelor's degree is the personification of an individual who presents higher levels of trust in AS.

This topic presents unique challenges, like the difficulty of manipulating risks in a laboratory environment and technology access. Fully autonomous traffic, for example, does not yet exist and research on how people behave towards this type of AS is nearly absent or simulated [9]. In general, our findings provide evidence that the propensity to take risks is a dispositional factor that affects dynamic human-agent trust formation and should be analyzed as a different aspect from the situational perception of risks.

7 Final Considerations

With the rapid advancement of artificial intelligence technologies and the increased interest from home goods, automobile, and tech industries, it can be expected that humans will have more opportunities to interact and collaborate with AS in our daily lives. Examining the effects of human-AS interactions on how trust is established and maintained is becoming an important and challenging priority. This research aims to shed light on the formation and calibration of trust in human and AS interactions, focusing on the influence of the propensity to take risks as a dispositional trust factor. By having a better understanding of individual differences and dispositional factors it will be possible to support the development of AS that are human-centered, which not only focuses on technology improvements such as system transparency but also reflects the individual propensity to take risk. Thus, personalized solutions to mitigating risks might depend on the capability to identify users' profiles that are most vulnerable in AS cybersecurity or other types of risks. Our findings provide information about the human factors that mediate human-AS interactions and thereby influence trust.

References

1. Cerrudo, C., Apa, L.: Hacking robots before skynet. IOActive Website, pp. 1–17 (2017)
2. Deb, S., Strawderman, L., Carruth, D.W., DuBien, J., Smith, B., Garrison, T.M.: Development and validation of a questionnaire to assess pedestrian receptivity toward fully autonomous vehicles. Transp. Res. Part C: Emerg. Technol. **84**, 178–195 (2017)
3. Dohmen, T., Falk, A., Huffman, D., Sunde, U., Schupp, J., Wagner, G.G.: Individual risk attitudes: measurement, determinants, and behavioral consequences. J. Eur. Econ. Assoc. **9**(3), 522–550 (2011)
4. Hillesheim, A.J., Rusnock, C.F., Bindewald, J.M., Miller, M.E.: Relationships between user demographics and user trust in an autonomous agent. In: Proceedings of the Human Factors and Ergonomics Society Annual Meeting, vol. 61, pp. 314–318. SAGE Publications, Los Angeles (2017)
5. Hoff, K.A., Bashir, M.: Trust in automation: integrating empirical evidence on factors that influence trust. Hum. Factors **57**(3), 407–434 (2015)
6. Hulse, L.M., Xie, H., Galea, E.R.: Perceptions of autonomous vehicles: relationships with road users, risk, gender and age. Saf. Sci. **102**, 1–13 (2018)
7. Lamb, K., Huang, H.Y., Marturano, A., Bashir, M.: Users' privacy perceptions about wearable technology: examining influence of personality, trust, and usability. In: Advances in Human Factors in Cybersecurity, pp. 55–68. Springer (2016)
8. Lee, J.D., See, K.A.: Trust in automation: designing for appropriate reliance. Hum. Factors **46**(1), 50–80 (2004)
9. Linkov, V., Zámečník, P., Havlíčková, D., Pai, C.W.: Human factors in the cybersecurity of autonomous cars: trends in current research. Front. Psychol. **10**, 995 (2019)
10. Lion, R., Meertens, R.M.: Seeking information about a risky medicine: effects of risk-taking tendency and accountability. J. Appl. Soc. Psychol. **31**(4), 778–795 (2001)
11. Lyons, J.B., Stokes, C.K.: Human–human reliance in the context of automation. Hum. Factors **54**(1), 112–121 (2012)
12. Madhavan, P., Wiegmann, D.A.: Similarities and differences between human–human and human–automation trust: an integrative review. Theor. Issues Ergon. Sci. **8**(4), 277–301 (2007)
13. Marsh, S., Dibben, M.R.: The role of trust in information science and technology. Ann. Rev. Inf. Sci. Technol. **37**(1), 465–498 (2003)
14. Meertens, R.M., Lion, R.: Measuring an individual's tendency to take risks: the risk propensity scale 1. J. Appl. Soc. Psychol. **38**(6), 1506–1520 (2008)
15. Pavlou, P.A.: Consumer acceptance of electronic commerce: integrating trust and risk with the technology acceptance model. Int. J. Electron. Commer. **7**(3), 101–134 (2003)
16. Perkins, L., Miller, J.E., Hashemi, A., Burns, G.: Designing for human-centered systems: situational risk as a factor of trust in automation. In: Proceedings of the Human Factors and Ergonomics Society Annual Meeting, vol. 54, pp. 2130–2134. SAGE Publications, Los Angeles (2010)
17. Singh, I.L., Molloy, R., Parasuraman, R.: Automation-induced "complacency": development of the complacency-potential rating scale. Int. J. Aviat. Psychol. **3**(2), 111–122 (1993)
18. Sitkin, S.B., Weingart, L.R.: Determinants of risky decision-making behavior: a test of the mediating role of risk perceptions and propensity. Acad. Manag. J. **38**(6), 1573–1592 (1995)
19. Strauss, W., Howe, N.: Generations: The History of America's Future, 1584 to 2069. William Morrow & Co., New York (1991)

The Impact of Fake News on the African-American Community

Wayne Patterson[1(✉)], Augustine Orgah[2], Suryadip Chakraborty[3], and Cynthia E. Winston-Proctor[4]

[1] Patterson and Associates, 201 Massachusetts Ave NE, Suite 316, Washington DC 20002, USA
waynep97@gmail.com

[2] Department of Physics and Computer Science, Xavier University of Louisiana, New Orleans, LA 70125, USA
aaorgah@xula.edu

[3] Department of Computer Science, Johnson C. Smith University, Charlotte, NC 28216, USA
schakraborty@jcsu.edu

[4] Department of Psychology, Howard University, Washington DC 20059, USA
cewinston@howard.edu

Abstract. An issue of increasing importance in the past few years has been what is generally referred to as "fake news". Although there is considerable evidence of such deceptive communication over many centuries, the sheer difference in deception techniques of such communication in an electronic environment has allowed the perpetrators the ability to disguise it in many forms that could not be seen in communication vehicles as print or electronic media such as radio or television. Techniques developed in the context of storable electronic information have allowed fake news items to take on a wider variety of disguises. In addition, with the access to electronic information being available in recent years to a large percentage of the world's population, the effect of such misleading information has had a much wider sphere of impact. As a consequence, many actors have developed sophisticated tools to convince even very diligent readers of the legitimacy of the false information purveyed. Many examples of this arose in the 2016 United States Presidential election. In particular, many items, supposedly from the Russian government, were aimed at reducing the African-American participation in that election. Our research attempted to assess the effectiveness of those attacks.

Keywords: Fake news · Historically Black Colleges and Universities · satire · COMPROP

1 Introduction

Perhaps the most recent and consequential examples of "fake news" stem from the 2016 United States Presidential Election. This has been discussed widely in many publications. In particular, we aimed to measure the impact of "fake news" on readers' susceptibility depending on the type of subject of such news. In particular, we aimed to measure

I. Corradini et al. (Eds.): AHFE 2020, AISC 1219, pp. 30–37, 2020.
https://doi.org/10.1007/978-3-030-52581-1_5

the impact of "fake news" on readers' susceptibility depending on the news subject type. It is now known from the Oxford University Computational Propaganda Research Project (COMPROP) that many posts were directed at reducing the participation of the African-American community in the 2016 US Presidential Election.

In our research, we have developed a test to try to determine "a susceptibility factor": to whether or not an audience of primarily African-American undergraduate students at two "Historically Black" universities (HBCUs) were likely to accept as a level of belief in a whole range of news through websites that might be considered truthful or "fake news".

For the test, we collected 25 news items which were either presumed to be true, were false items of a general character, and false items specifically aimed at the African-American community. Through a number of measures, we determined the level of believability in each item, and also the level of intensity of feeling in each item as well.

For one-half of the test participants chosen at random, a session on detection of "fake news" developed by the authors prior to the participants taking the test; the other participants were not provided with such instruction.

In this paper, we report on the results of our findings in this regard, both in terms of the levels of believability and intensity of feeling for all participants, as well as any differences that can be attributed to the prior instruction.

2 The Impact of Fake News on African-American Voters

There has been a substantial rise in recent years in the propagation of false information intended to confuse persons in a target environment, a methodology that has been normally referred to as "fake news".

This has become a global phenomenon, and perhaps the leading analysis of this phenomena has been developed as part of the Oxford University Computational Propaganda Research Project. [1]

In our case, we have attempted to develop further analysis of the impact of such "fake news" on a specific target population, the African-American population which had been subjected to attacks of such misinformation, particularly in the lead up to the 2016 United States Presidential election.

In order to try to determine the impact of such a strategy, we have surveyed an audience of primarily African-American students at two HBCUs (or Historically Black Colleges and Universities). By providing a series of test items, we tried to measure the impact of various types of information provided to such an audience and measure their reaction to such items.

The participants were provided with a set of 25 news or information items, which in all fell into three categories: (a) presumably true or real items (R); (b) false items that could be considered "fake news" (F), many specifically aimed at the African-American community; and (c) false items that could be considered as satire (S), or items specifically labeled as being false.

The reactions that participants were asked to provide were based on an intensity level, using a Likert scale from 1 to 5 with respect to the following types of reaction: anger, disbelief, information level, agreement, satisfaction, incredulity, or believability.

3 Research Design

The items in Table 1 below constituted the subject matter for the test. These items were provided to the participants in one of two formats: in some cases, as the relevant URL for the participants to view online, and in other cases, as a screen shot or photocopy of the relevant page. The 25 items are drawn from many sources, as indicated by their heading or title, and their Web site or other location. Also, a number of the items were identified as having content particularly aimed at the African-American community (AA Oriented).

Table 1. Test items.

Item	Heading or title	Web or other site	Category	AA oriented
1	"Florida Millionaire Arrested after Authorities Discover Over 700 Bodies"	EmpireNews.net	Fake	
2	"Amazon Releases Crazy, Trash-talking Speaker"	CBSnews.com	Real	
3	"Never Forget Eric Garner"	comprop.oii.ox.ac.uk	Fake	Yes
4	"Uber Controversy: How Uber has Partnered with Mafia"	humoropedia.com	Satire	
5	"Invasion Begins! Migrant Caravan Arrives at US/Mexico Board"	infowars.com	Fake	
6	"Mitt Romney Considering Dropping out, Buying Canada Instead"	currantdaily.com	Satire	
7	"Washington State Highway Closed after Cars get Trapped in Tumbleweeds"	huffpost.com	Real	
8	"I Won't Vote. Will You?"	BlackMatters facebook (closed)	Fake	Yes
9	"Mozart Was Black."	i.stack.imgur.com	Fake	Yes
10	"Vegetarianism More Dangerous Than Smoking"	currantdaily.com	Satire	
11	"What's 50 Cent's phone number? The weird inquiries made to the Foreign Office"	news.sky.com	Real	Yes
12	"Media Alert: Jack Burkman, Jacob Wohl present bombshell witness/evidence"	Twitter.com	Fake	

(*continued*)

Table 1. *(continued)*

Item	Heading or title	Web or other site	Category	AA oriented
13	"Before You Vote-Listen to MJ"	BlackMatters facebook (closed)	Fake	Yes
14	"Wakanda free trade forever? Fictional nation removed from US trade list"	reuters.com	Real	Yes
15	"5 of the least ethical medical experiments in modern history"	cracked.com	Satire	
16	"Random: what's your talent??? Me: I'm black."	me.me	Fake	Yes
17	"Washington State Highway Closed at after Cars get Trapped in Tumbleweeds"	huffpost.com	Real	
18	"All we have to do is stand up and their little game is over"	democraticunderground.com	Fake	Yes
19	"Amazon Releases Crazy, Trash-talking Speaker"	humoropedia.com	Satire	
20	"A Bull has an afternoon boat in Baltimore"	npr.org	Real	Yes
21	"The stories from 2019 YouTube hopes you forgot"	cracked.com	Satire	
22	"Zombie Preparedness"	cdc.gov	Fake	
23	"Black College Students are not excited this election, think Hillary's a liar"	dailycaller.com	Fake	Yes
24	"Thief hijacks band of lobsters, crashes into another lobster truck"	nypost.com	Real	
25	"Trump Warns that Florida recount could set dangerous precedent a person with the most votes winning"	newyorker.com	Fake	

It should be noted that the Oxford University COMPROP identified many efforts of the Russian government, through a government entity identified as "IRA" or Internet Research Agency. We have used examples Oxford has identified as having been produced by this IRA in items 3, 8, 9, 13, 16, 18 and 23 above. In the text below, when we mention "Russia" it is in particular reference to this "IRA".

These test items were given to 47 test participants at the participating universities, Xavier University of Louisiana in New Orleans and Johnson C. Smith University in Charlotte, North Carolina. The participants were asked to classify each of the 25 items as fake news (F), real (R), or satire (S). In addition, a subset of the test participants was also given preparation in terms of a document entitled "Training Module for Fake News Detection" [2].

The results of the survey were analyzed with respect to several subsets as described in Table 2 below.

Table 2. Respondents to test items.

Respondents	n	% CORRECT CHOOSING F	% CORRECT CHOOSING R	% CORRECT CHOOSING S
All	47	34.0%	38.0%	29.8%
African-Americans (AA)	41	33.5%	34.8%	29.7%
Others	6	26.4%	59.5%	30.6%
Persons receiving Training Module	12	42.4%	44.0%	34.7%
Not receiving Training Module	35	31.2%	35.9%	28.1%
Reviewed URL when responding	24	35.8%	42.3%	34.0%
Reviewed screenshot	23	32.2%	33.5%	25.4%
Subjects with African-American content	47	28.6%	47.3%	29.3%
African-American responders to AA content	41	29.0%	47.4%	30.1%
Other responders to AA content	6	25.9%	46.3%	24.1%

4 Analysis of Results of F-R-S Choice

There are several interesting conclusions that can be drawn from the results in the above table. First, the selection of certain of the respondents to receive prior instruction based on techniques for discerning that publications may constitute "fake news" had an effect on the overall ability of the respondents to correctly identify the nature of the test items. Perhaps most important, though the participants who received this prior instruction were able to determine the "fake news" items 8.4% more often than the entire group, and 11.2%

better than those who had not received such prior instruction notes. In addition, this also enabled the respondents with those notes to improve their score on recognizing both real news items and satire between 5 and 6 per cent.

A second division of the participants, also shown in Table 2 depended on whether they received the test items electronically as a URL or by viewing a screenshot, and the URL version was slightly better in every case, and almost 10% better in the case of recognizing the real items and the satire.

There are several interesting conclusions that can be drawn from the results in the above table. First the selection of certain of the respondents to receive prior instruction based on techniques for discerning that publications may constitute "fake news" had an effect on the overall ability of the respondents to correctly identify the nature of the test items. Perhaps most important, though the participants who received this breathing were able to determine the "fake news" items 8.4% for often than the entire group, and 11.2% better than those who had not received such prior instruction. In addition, this also enabled the respondents with those notes to improve their score on recognizing both real news items and satire 5 between 5 and 6%.

5 Analysis of Study of "Impressions"

There were a number of interesting conclusions that could be drawn from the respondents weighting of answers concerning their impressions or feelings upon conducting the survey. These responses were analyzed with respect to the Likert scale, with particular attention to the predominance of answers that tended to side with agreement on each of the topics.

The specific use of the Likert scale (1 to 5) is described in Table 3 below.

Table 3. Use of the Likert scale.

Level of agreement	Scale
Strongest agreement	5
Strong agreement	4
Moderate agreement or disagreement	3
Somewhat disagree	2
Strongly disagree	1

As indicated above, for each test item 1–25, the participants were asked to provide their impressions on a number of categories, based on an intensity level, using the Likert scale described in Table 3 with respect to the following types of reaction: anger, disbelief, information level, agreement, satisfaction, incredulity, or believability.

The test items that registered the strongest level of reaction (averaging above 2.5 on the average of Likert scale values are described in Table 4.

Table 4. Respondents to test items.

Type of reaction	Items > 2.5 as % of all responses	Test Items, ranked from highest value to lowest value above 2.5
Anger	12%	3, 8, 5
Disbelief	60%	1, 4, 10, 9, 14, 24, 11, 5, 19, 6, 12, 7, 8, 13, 22
Informative	16%	3, 15, 18, 2
Agreement	16%	3, 18, 15, 13
Satisfaction	4%	3
Incredulity	4%	1
Believable	32%	3, 18, 2, 15, 20, 21, 25, 23

One particular case in point was item 3 on the test, the "Eric Garner" item, which raised the highest levels of anger, informative, agreement, satisfaction, and believability. This particular item has been classified as "fake news" as part of the Russian-based propaganda, nevertheless it seems to have had a greater impact.

It should be noted that although the Eric Garner entry is fake, the actual death of Eric Garner was widely discussed and reported as Mr. Gardner was killed in Brooklyn by being choked to death in 2014 by New York City police officers [3].

Of the items that seemed to register the greatest anger in the respondents were the Garner item (3), another Russian propaganda item "I won't vote. Will you" (8) and item 5 "Invasion begins" on illegal migration, not particularly an African-American issue, but one related to other cultures, yet perhaps appealing to a minority audience. It should be noted that all three items 3, 5, 8 are considered "fake news".

Although 15 of the 25 items (60%) registered very high in Disbelief, it should be noted that the highest ranking was indeed a false item, "Florida Millionaire Arrested after Authorities Discover Over 700 Bodies" (1). Less than half of the 15 were actually false (7/15) and 4 were satire and 4 were real.

Of the 4 items that were considered highly Informative, including the Eric Garner item was another Russian-based faith item "All we have to do is stand up" (18); and a humorous item about an Amazon device (2).

Items that registered high degrees of Agreement were led by the Eric Garner item, but also the fake item "All we have to do is stand up" (18) and the Michael Jackson item "Before You Vote-Listen to MJ" (13) and "Random: what's your talent??? Me: I'm black." (16). All of these were false.

Eight items registered above average in terms of Believability. Half were fake and one-quarter each were real and satire.

"Zombie Preparedness": One interesting set of responses arose from test item (22) on "Zombie Preparedness", which actually appears in a United States government website, for the Centers for Disease Control, www.cdc.gov. Although the meaning of this government website was probably satirical, 12.7% of participants felt this item was real, 55.3% felt it was fake, and only 31.9% that it was satire.

6 Conclusions

There is significant evidence that items of fake news which were directed in part as a deliberate disinformation campaign by Russian interference in the 2016 election would have had a significant impact on African-American college students at two Historically Black Universities. In particular, a few of these items registered highest on the scales of raising anger engendering agreement, registering satisfaction, and being believable. This underlines the prior research that has been conducted on the impact of what has been called "fake news". However, on the other hand, our research has shown that the ability to mitigate these results can be done if the consumers of such fake news are given prior training in how to observe what is read from a political point of view, as was done with the subset of the test participants and reported in Sect. 3 above.

We also wish in future work to extend this exploratory study to a larger scale multisite study.

References

1. Howard, P.N., Ganesh, B., Liotsiou, D.: The IRA, Social Media and Political Polarization in the United States, 2012-2018, Computational Propaganda Research Project. University of Oxford, Oxford (2018)
2. Patterson, W., Winston-Proctor, C.: Behavioral Cybersecurity (Chapter 24). CRC Press, Orlando (2019)
3. Goodman, J.D.: Eric Garner Died in a Police Chokehold. Why Has the Inquiry Taken So Long? New York Times, November 7, 2018

Cybersecurity Tools and Analytics

Detecting Identity Deception in Online Context: A Practical Approach Based on Keystroke Dynamics

Matteo Cardaioli[1,2](✉), Merylin Monaro[3], Giuseppe Sartori[3], and Mauro Conti[1]

[1] Department of Mathematics, University of Padova, via Trieste 63, 35131 Padua, Italy
matteo.cardaioli@phd.unipd.it, mauro.conti@unipd.it
[2] GFT Italy, via Sile, 18, 20139 Milan, Italy
[3] Department of General Psychology, University of Padova, via Venezia 8, 35131 Padua, Italy
{merylin.monaro,giuseppe.sartori}@unipd.it

Abstract. Keystroke dynamics has been recently proved to be an effective behavioral measure to detect subjects who provide false demographic information in online contexts. However, current techniques still suffer from some limits that restrict their practical application, such as the use of errors as a key feature to train the lie detectors and the absence of normalized features. Here, an extension of a keystroke dynamics technique, which was recently proposed to detect faked identities, is reported with the goal to overcome these limitations. Using a Quadratic Discriminant Analysis an accuracy up to 92% in the identification of faked identities has been reached, even if errors were excluded from predictors and normalized features were included. The classification model performs similarly to those previously proposed, with a slightly lower accuracy (−3%) but overcoming their important practical limitations.

Keywords: Keystroke dynamics · Lie detection · Identity verification

1 Introduction

In the current historical and cultural framework, identity verification is an increasingly urgent problem. Faked identities are used for a wide range of criminal purposes, both in the real word and in the internet environment [1]. Concerning online security, the scenario becomes very intricate: identity alteration is common in social networks profiles [2] and often used with malicious intents (e.g., child grooming); identity thefts and frauds are often means to perpetuate illegal business and financial crimes. To deal with this issue, the research focused on techniques for identity verification that are based on the study of human behavior and, in particular, the study of the interaction between the user and the computer [3]. One of the most investigated measures is keystroke dynamics. It gives information about the typing pattern of the user who is engaged in typing a text on the keyboard [4].

I. Corradini et al. (Eds.): AHFE 2020, AISC 1219, pp. 41–48, 2020.
https://doi.org/10.1007/978-3-030-52581-1_6

Keystroke dynamics has been widely applied to the problem of user authentication or identification as a behavioral biometric measure [5]. However, similarly to other biometric techniques (e.g., fingerprint, face recognition) [6], it necessarily requires a high level of knowledge about the user, as a specific training is needed to distinguish the rightful user from the intruder based on their typing pattern [7]. Consequently, keystroke dynamics as a biometric method cannot be used to spot faked identities in the absence of ground truth.

More recent studies proposed to use keystroke dynamics as a lie detection technique [8], to spot people who self-declare false demographic information studying their keystroke pattern, but without any previous knowledge about the user [9–11]. One of these techniques has been proposed by Monaro et al. [10, 11] and consists of analyzing the keystroke dynamics while the user is engaged in compiling a form asking for control (e.g., gender) expected (e.g. name, surname, date of birth) and unexpected identity information (e.g., zip code, zodiac) [10, 11]. Control information is details about the subjects which are easily verifiable, such as gender or ethnicity. Expected information includes all the identity details which are commonly reported in the ID card or frequently asked by online forms (e.g., to subscribe social networks, for the online banking authentication). Unexpected information is details about the subjects' identity which are not usually asked, so the user is not prepared to provide them, such as the zodiac [12]. In other words, liars cannot prepare the responses to these questions in advance. The use of unexpected questions has proven to be an effective strategy to increase the probability of identifying liars [13]. Indeed, the liar needs to fabricate the fake response in real-time, checking the congruency of the response with the other faked information and maintaining credibility and consistency. This mental process results in an increment of cognitive load and, consequently, at the typing level, in an increment of response time, writing time and number of errors [11]. Based on these keystroke features, which were collected while the subjects responded to control, expected and unexpected questions, Monaro et al. have trained different machine learning classification models, reaching an accuracy of 95% in detecting users who compiled the form providing false demographic information [11].

Despite the excellent results, the technique proposed by Monaro et al. [10, 11] suffers from a limitation: the accuracy of the detection algorithm is mainly based on errors, that is the number of unexpected information for which the user does not know the answer. The practical limitation is twofold: first, in the realistic application of the technique, the user can simply avoid errors by searching the correct response on the internet. This increases the overall response time, making the temporal features (e.g., the average typing speed, the time between the presentation of the question and the first key pressed, etc.) more central. Secondly, to calculate errors, a software that knows the correct response to all possible questions is needed (e.g., the zip code of Tuscaloosa, Alabama State), which is very expensive in computational terms. Moreover, the above mentioned works [10, 11] do not take into account the individual differences in the typing pattern. For example, people who are used to writing with a computer keyboard every day in their job are basically more skilled than people who are not very familiar with the computer. In other words, the users' performance in typing faked or truthful demographic information is not compared with their keystroke baseline.

In this paper, we propose an extension of the keystroke dynamics technique to detect faked identities proposed by Monaro et al. [10, 11], but avoiding to use errors as a predictor of deception. Moreover, we performed a new feature extraction to obtain normalized measures, which are independent from the users' typing skills, and improve the generalization of the classification models.

2 Dataset

The two open accessed datasets collected and released by Monaro et al. in [11] were analyzed. The first dataset includes 60 participants who were recruited and performed the experiment in the laboratories of the Department of General Psychology of the University of Padova. The second dataset includes 145 participants who were recruited online. Participants were instructed to respond truthfully or lying to control (n = 4), expected (n = 8) and unexpected (n = 6) questions about their identity. Questions were presented in the central part of the computer screen and the subject was asked to type the response in a box, pressing ENTER to confirm the response and to pass to the next question. Truth-tellers were instructed to complete the experiment providing their real demographic information. Liars were asked, before starting the experiment, to learn a fake identity from a false ID card and, afterward, to complete the experiment pretending to be the person in the ID. For further details about the data collection procedure, see [11].

The keystroke features collected by the authors [11] were the following: number of errors (the number of fields for which incorrect information was entered), prompted-firstdigit (the interval between the onset of the question on the computer screen and the first key pressed), prompted-firstdigit adjusted GULPEASE (it is the prompted-firstdigit adjusted using the GULPEASE Index, a readability index for the Italian language), prompted-enter (the total time from the stimulus onset to ENTER), firstdigit-enter (the time between the first key pressed and ENTER), time before enter key down (the time between last key pressed and ENTER), answer length (the number of characters of the response), writing time (the typing speed calculated dividing the firstdigit-enter for the number of typed characters), down time (the timestamp for pressing each key), up time (the timestamp for releasing each key), up and down time (the sum of down time and up time for each key), press time (the duration between each key down and each key up), flight time (the interleaving time between each key up and the next key down), di-graphs (the sum of up time, down time or up and down time for two consecutive keys), tri-graphs (to the sum of up time, down time or up and down time for three consecutive keys), frequency of use of Shift, Del, Canc, Space and Arrows. For a more detailed description of the keystroke features, see [11].

3 Methods

We identified eight new keystroke features in addition to those suggested in [11]. The error rate has been completely excluded from the new set of features, overcoming the limitation of the previous studies [10, 11]. Moreover, in order to minimize the data noise due to the inter-participants differences, all the features have been normalized considering the typing baseline of each user. Finally, different machine learning algorithms have been trained using a leave-one-out cross validation technique.

3.1 Feature Extraction

As suggested in [13], we performed a preliminary feature selection based on the maximum correlation between the predictors and the dependent variable. Moreover, we excluded the error rate and we removed those predictors that showed an inter-correlation value higher than 0.85. Three features resulted to fit these constraints. In particular: prompted-firstdigit adjusted GULPEASE, firstdigit-enter and, writing time. We grouped these features by expected and unexpected questions, then, we calculated the average and the SD of each group. Finally, for each feature, we subtracted the average and SD of the expected group respectively from the average and SD of the unexpected group. We obtained six new normalized features: delta mean prompted-firstdigit adjusted GULPEASE (DMG), delta mean firstdigit-enter (DME), delta mean writing time (DMW), delta SD prompted-firstdigit adjusted GULPEASE (DSDG), delta SD firstdigit-enter (DSDE) and, delta SD writing time (DSDW). We considered all the possible feature combinations that contains at least two predictors, obtaining in total 58 possible cases (i.e. $2^6 - 6$).

3.2 Learning Algorithms

For the prediction, we decided to evaluate four different ML classifiers. In particular:

1. Non-linear SVM with Gaussian Kernel. An SVM is an ML model for binary classification. The algorithm calculates the best hyperplane that separates the points belonging to one class to the point belonging to the other. The Gaussian Kernel is applied to patterns that are not linearly separable and maps the original points into a new space.
2. Naive Bayes. It is a probabilistic classifier based on the Bayes theorem, with the assumption that the predictors are independent. In training, it estimates the probability distribution of the predictors, while in testing it calculates the posterior probability of the data. The classification is made on the base of the higher posterior probability.
3. Random Forest. It is an ensemble of decision trees. The classification is made from each tree prediction. The class that receives more predictions from the decision trees is chosen as the Random Forest predicted class.
4. Quadratic Discriminant. It is based on the assumption that the data of each class derived from a normal distribution. In training, the algorithm estimates the Gaussian distribution parameters of each class. In test, the trained algorithm selects the class with the smallest misclassification cost.

3.3 Models Evaluation

To obtain robust models and to generalize as much as possible our results, we performed leave-one-out cross-validation. This approach is similar to cross-validation but the number of folds is equal to the number of instances in the dataset [14]. Thus, the test set is composed of only one instance and the training set is composed of all the other instances. We iteratively applied the leave-one-out cross-validation to both offline and online datasets and trained our models with 5-fold cross-validation. To improve the accuracy and the generalization of our algorithms, we performed a feature selection. For each of the four algorithms presented in Sect. 3.2, we evaluated all the possible combination of features presented in Sect. 3.1 and selected the set of predictors that maximized the validation accuracy of every fold of the leave-one-out cross-validation. Finally, for each ML algorithm, we calculated the global classification accuracy as the average of all the test sets predictions (i.e. 60 for the offline dataset and 145 for online dataset). A summary diagram of the process is presented in Fig. 1.

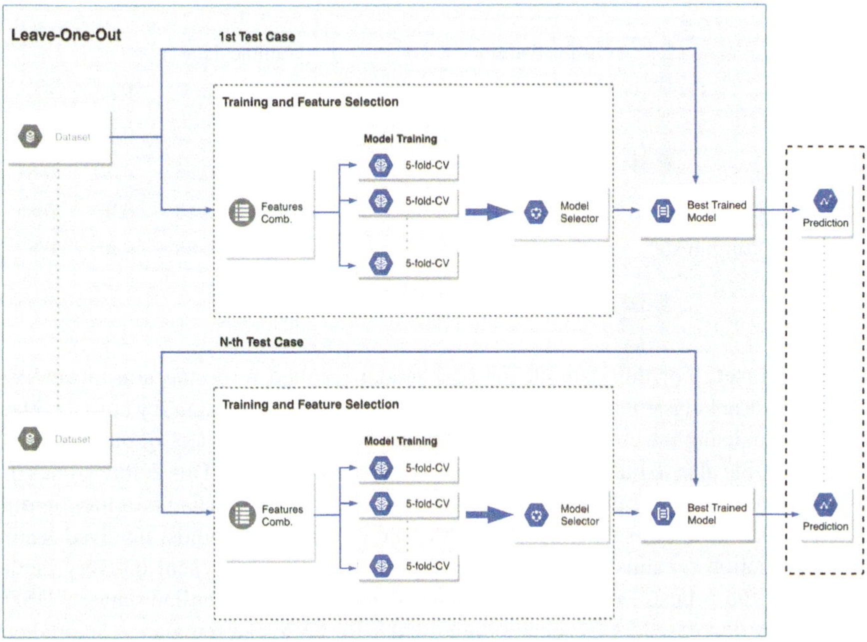

Fig. 1. Summary of the whole process, applied for all the four SVM, Naive Bayes, Random Forest and Quadratic Discriminant algorithms

4 Results and Discussion

In this section, we discuss the results obtained by our models for both the offline and online datasets, analyzing the role of the individual features in the prediction of liars and truth-tellers behaviors.

Offline Dataset. In Table 1, we report the prediction accuracy of each classifier on the test set. For this dataset, we are able to achieve 92% accuracy with Quadratic Discriminant. In particular, the accuracy obtained with this algorithm is comparable with the best accuracy obtained by Monaro et al. [13], but without using the error rate feature. Moreover, our models outperform those trained without the error rate feature presented in [13], improving accuracy by more than 20%. This demonstrates also that the new set of feature significantly improves prediction performance. To understand which features are most involved in the prediction, we calculated the frequency of each selected set of features in validation (see Fig. 2). The results show that DMG is always used as a feature for all the models and the second most used is DSDG (80% on average), except for the Naive Bayes. The most used feature combination resulted to be DMG, DME, DSDG and, DSDW.

Table 1. Accuracy (Acc) results of our models for the offline and online datasets. In the table are reported also precision (Prec), recall (Rec) and, F1 score (F1).

Model	Offline dataset				Online dataset			
	Acc	Prec	Rec	F1	Acc	Prec	Rec	F1
SVM	0.87	0.82	0.93	0.88	**0.90**	0.93	0.88	0.90
Naive Bayes	0.87	0.89	0.83	0.86	0.81	0.95	0.70	0.81
Random Forest	0.85	0.84	0.87	0.85	0.86	0.88	0.88	0.88
Quadratic Discriminant	**0.92**	0.90	0.93	0.92	0.85	0.88	0.85	0.87

Online Dataset. For this dataset, the best model resulted to be the non-linear SVM Gaussian Kernel classifier, see Table 1. This model achieves an accuracy comparable to [13], without using the error rate feature. Moreover, the performances of our algorithms are comparable also to our results obtained for the online dataset. This demonstrates that our algorithms can generalize well the problem of classifying participants as liars or truth-tellers also in different scenarios. As for the online dataset, the most involved feature in the prediction remains the DMG (see Fig. 2). It is always present in every model, confirming the robustness of this parameter. Compared to the online dataset, DSDG significantly decreased its frequency in particular for SVM but remains a top predictor for the Quadratic Discriminant. DME, DMW, DSDG and, DSDE show a similar overall frequency, respectively 68%, 60%, 66% and, 63%. The most used feature combination resulted to be DMG, DME, DSDG and, DSDE, differing to the online dataset by only the last feature.

To conclude, we showed that excluding errors as predictors it is still possible to reach a good accuracy in detect deception based on the keystroke pattern. Indeed, our model performs similar to those proposed by [13] with a slightly lower accuracy, but overcoming the important limitation of calculating errors, both for offline and online datasets.

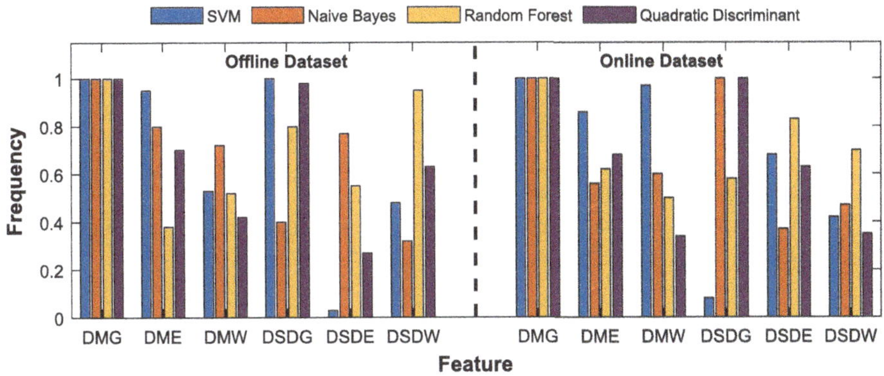

Fig. 2. Frequency of predictors after feature selection process for both offline and online dataset.

References

1. Cano, A.E., Fernandez, M., Alani, H.: Detecting child grooming behaviour patterns on social media. In: Aiello, L.M., McFarland, D. (eds.) SocInfo 2014. LNCS, pp. 412–427. Springer, Cham (2014)
2. Gross, R., Acquisti, A.: Information revelation and privacy in online social networks. In: Proceedings of the 2005 ACM Workshop on Privacy in the Electronic Society, pp. 71–80 (2005)
3. Moskovitch, R., Feher, C., Messerman, A., Kirschnick, N., Mustafic, T., Camtepe, A., Löhlein, B., Heister, U., Möller, S., Rokach, L., Elovici, Y.: Identity theft, computers and behavioral biometrics. In: 2009 IEEE International Conference on Intelligence and Security Informatics (InISI 2009) (2009)
4. Ahmad, N., Szymkowiak, A., Campbell, P.A.: Keystroke dynamics in the pre-touchscreen era. Front. Hum. Neurosci. **7**, 835 (2013). https://doi.org/10.3389/fnhum.2013.00835
5. Monrose, F., Rubin, A.D.: Keystroke dynamics as a biometric for authentication. Future Gener. Comput. Syst. **16**, 351–359 (2000). https://doi.org/10.1016/S0167-739X(99)00059-X
6. Spolaor, R., Li, Q., Monaro, M., Conti, M., Gamberini, L., Sartori, G.: Biometric authentication methods on smartphones: a survey. PsychNology J. **14**, 87–98 (2016)
7. Teh, P.S., Teoh, A.B., Yue, S.: A survey of keystroke dynamics biometrics. Sci. World J. (2013). https://doi.org/10.1155/2013/408280
8. Grimes, G., Jenkins, J.L., Valacich, J.S.: Assessing credibility by monitoring changes in typing behavior: the keystrokes dynamics deception detection model. In: Hawaii International Conference on System Sciences, Deception Detection Symposium (2013)
9. Monaro, M., Spolaor, R., QianQian, L., Conti, M., Gamberini, L., Sartori, G.: Type me the truth!: Detecting deceitful users via keystroke dynamics. In: Proceedings of the 12th International Conference on Availability, Reliability and Security, ARES 2017, Reggio Calabria, Italy (2017)
10. Monaro, M., Businaro, M., Spolaor, R., Li, Q.Q., Conti, M., Gamberini, L., Sartori, G.: The online identity detection via keyboard dynamics. In: Arai, K., Bhatia, R., Kapoor, S. (eds.) FTC 2018. AISC, vol. 881, pp. 342–357. Springer, Cham (2019). https://doi.org/10.1007/978-3-030-02683-7_24
11. Monaro, M., Galante, C., Spolaor, R., Li, Q.Q., Gamberini, L., Conti, M., Sartori, G.: Covert lie detection using keyboard dynamics. Sci. Rep. **8**, 1976 (2018). https://doi.org/10.1038/s41598-018-20462-6

12. Monaro, M., Gamberini, L., Sartori, G.: Identity verification using a kinematic memory detection technique. In: Hale, K., Stanney, K. (eds.) Advances in Neuroergonomics and Cognitive Engineering. Advances in Intelligent Systems and Computing, vol. 488, pp. 123–132. Springer, Cham (2017)
13. Vrij, A., Leal, S., Granhag, P.A., Mann, S., Fisher, R.P., Hillman, J., Sperry, K.: Outsmarting the liars: the benefit of asking unanticipated questions. Law Hum Behav. **33**, 159–166 (2009). https://doi.org/10.1007/s10979-008-9143-y
14. Sammut, C., Webb, G.: Encyclopedia of Machine Learning and Data Mining. Springer, US (2017)

An Analysis of Phishing Emails and How the Human Vulnerabilities are Exploited

Tanusree Sharma[1] and Masooda Bashir[2(✉)]

[1] Illinois Informatics Institute,
University of Illinois at Urbana-Champaign, Champaign, IL 61820, USA
tsharma6@illinois.edu
[2] School of Information Sciences,
University of Illinois at Urbana-Champaign, Champaign, IL 61820, USA
mnb@illinois.edu

Abstract. Humans continue to be considered as the weakest link in securing systems. While there are a variety of sophisticated system attacks, phishing emails continues to be successful in gaining users attention and leading to disastrous security consequences. In designing strategies to protect users from fraudulent phishing emails, system designers need to know which attack approaches and type of content seems to exploit human limitations and vulnerabilities. In this study, we are focusing on the attackers' footprints (emails) and examining the phishing email content and characteristics utilizing publicly available phishing attack repository databases. We analyzed several variables to gain a better understanding of the techniques and language used in these emails to capture users' attention. Our findings reveal that the words primarily used in these emails are targeting users' emotional tendencies and triggers to apply their attacks. In addition, attackers employ user-targeted words and subjects that exploits certain emotional triggers such as fear and anticipation. We believe our human centered study and findings is a critical step forward towards improving detection and training programs to decrease phishing attacks and to promote the inclusion of human factors in securing systems.

Keywords: Human factors · Phishing email · Cybersecurity · Emotion · Psychology

1 Introduction

Humans are often considered the weakest link in cybersecurity. With the help of Social engineering, attackers are exploiting many human tendencies to persuade them to have access to the system. Social engineering is the psychological manipulation of people to make them give up their confidential information [1]. The increasing amount of technological communication tools (email, text messaging) are paving the way for attackers to grab users' attention in different means and at the same time creating new threat landscapes for social engineering attacks. Nowadays, attacks by utilizing social engineering

I. Corradini et al. (Eds.): AHFE 2020, AISC 1219, pp. 49–55, 2020.
https://doi.org/10.1007/978-3-030-52581-1_7

are one of the advanced persistent threats by exploiting a higher amounts of users' personally identifiable information. One such example is phishing emails which are the consistent and successful attacks that any organization is facing over time. Phishing is the attempt to access individuals' systems to obtain information, for example, usernames, passwords, and credit card details and so on and it usually is carried out by email spoofing and distribution of email with noticeable contents and topics [2, 3]. This is one of the most common and frequent cyberattacks that leave a great effect (both economically and reputation) on any organization and government. The practice of phishing was originated via AOL floppy disk when individuals were not willing to pay for internet access and used alternative of thirty days free trial and is termed as "phishing" on the 90's [4]. Some found a way to change their screen name as if they were AOL administrators and were able to "phish" for log-in credentials to continue accessing the Internet for free [4]. By definition, "phishing" is an attempt, via message and email to catch the attention of computer users to reveal sensitive personal information such as passwords, date of birth, credit cards, and social security numbers and so on.

Phishing attacks can be proved dangerous for an organization if their employees get trapped by legitimate-looking phishing emails. There are several countermeasures that are developed to prevent and minimize phishing attacks and its harmful consequences such as risk of losing money/personal information and reputation. For example, security awareness training programs, detection models, and so on. While there are studies in effective countermeasures through security awareness training programs and detection models [5, 6] much of the research focuses on building a system that can detect phishing emails by senders' domain names, associated email addresses, and URLs to determine maliciousness. However, phishing attackers are continuing to generate malicious emails that are successful in catching users' attention using innovative ways that are more sophisticated and preys on the human susceptibilities. While state-of-the-art cybersecurity countermeasures are taken at the system level to prevent and filter phishing emails, there are very few studies that focus on the patterns, evolution of attack types and users' emotional elements associated with phishing emails. In particular, an analysis of the phishing email content is a critical step towards decreasing phishing attacks that prey on human vulnerabilities.

In our study, we have studied through a knowledge-based approach (lexicon and context) to extract and reveal different forms of historical phishing attack scenarios and map the intention behind those. We believe this type of investigation can provide insight on attack patterns and therefore, can help in developing security measures that are appropriate.

2 Background

Social Engineering, in particular, Phishing attacks is a field of study that grabs the attention of many researchers in different fields and various systematic studies. A significant number of researchers have investigated: URL based approaches, server-client-side approaches, developing effective training programs, designing tools to detect phishing in the system, and so on. For example, Google and Microsoft's effectiveness of the blacklists has been tested by the Phishing Detection Schemes approach to finding out the efficiency [7]. In recent years, increasingly there are other approaches such as automated analysis and detection utilizing machine learning to detect phishing email websites and other security related measures [8–11, 20]. Furthermore, Web page Content analysis for phishing detection is a content-based approach to identify probable phishing websites by storing and mining data [12–15]. There are even research studies that have conducted hybrid approach of adopting URL based and web content-based approaches. For example, a research team from Google accomplished a large-scale automatic classification of phishing web pages [16] by analyzing both the URL and the content of the web pages. URL analysis and phishing web content have been becoming in the spotlight to highlight the ways of detecting phishing in electronic communication.

From the existing research, we can see that different approaches have already been explored/employed to detect phishing emails through URL analysis, webpage content analysis, phishing detection scheme, email content and other such mechanisms. Despite all of these sophisticated technical methods to combat phishing email, attackers are still finding their way to get into our inboxes/systems. Since phishing attacks typically target users and take advantage of their human vulnerabilities there are typically two types of methods used to increase detection: the automatic software detection and user awareness. Most of the above research reviewed is based on highly technical perspective that is to build and design a detection system. Our aim in this study is to address human factors and human psychology behind the phishing attacks. Thus, we examined both users and attackers to determine the most frequent type of phishing emails and to identify what type of emotional plea is related to those patterns. We believe, our heuristic approach will unfold initial critical details in order to form a rule-based test to create a mapping of incoming phishing attacks.

3 Methodology

This section describes the process of screening and collecting phishing email samples from two public phishing archives.

3.1 Data Collection

Our study mainly focused on content analysis of reported phishing emails from the selected publicly available databases. Our selected 2 public databases are: "Berkeley Information Security Office" and "SecureIT-Kent State University". Our study is designed to conduct a content analysis of phishing emails with subject, contents, date, time, compelling word and other information to gain a better understanding of the techniques and language that have been used to catch users' attention. Our collected total phishing emails: N = 217. We retrieved the phishing email data from Berkeley Information Security Office and SecureIT-Kent State University from the time interval 2015 to 2019.

3.2 Coding Strategy

In order to prepare the retrieved data for further analysis, we performed the following coding steps.

General Characteristics. Basic information was captured from the selected archives which included Email subject; Email content (body); sending date, time and day; Compelling words; presence of sender name and email (in Y/Yes or N/No format), used email signatures, used legitimate logo; recipients' email (Blind or Non-Blind format).

Subjects in Phishing Emails. To better analyze the phishing email content, we developed/categorized 10 types of subjects that attackers use to send phishing emails (1) Online account (update, upgrade, verification, notification and so on), (2) Payment/transaction, (3) Document shared, (4) Tax/payroll, (5) Job hiring/business, (6) Shipment, package delivery, (7) Audio message, (8) Education, (9) Donation and (10) Others. Figure 1 shows the initial coding results and distribution of emails within each of the 10 subject types.

4 Results

As stated above our goal in this study was to better understand historical attack types and how they are related to human emotional vulnerabilities. Thus, we employed different emotional plea types detection utilizing the integration of information from email body, subject and most frequently used words.

We conducted our main content analyses approach was utilizing knowledge-based techniques (in many cases, referred to as lexicon-based techniques) [18], utilize domain knowledge to detect particular emotional plea types that present in content.

Sample Description and General Characteristics. Our overall email collection was N = 215. All of the emails are from the publicly available databases. All of those emails are classified in 10 Subjects. Based on knowledge-based approaches and depending on the context of the email, we were able to calculate the frequency of the emails in each subject. The most frequently used Subject of capturing users' attention is through online account verification, update, confirmation, validation aspect which is 23.26% of the total amount of collection, see Fig. 1. The second most frequent is Shared document where users are sent a different kind of attachment via google drive, doc, dropbox with the URL link to open which is arguably the most efficient way to get users to click a link which makes up the 18.6% of the emails.

From our collected data and analysis, it appears that users usually open emails if it is regarding a payment, transaction or bank related issues which leads to the 3rd most frequent (17.2%) Subject type of phishing emails followed by Audio call (9.77%) with URL link job/business posting and then (8.37%), Education and school-related topics. Additional Subjects included, urgent notification from admin, HR, faculty, chancellor and so on (6.51%), shipment/order/package delivery mostly with Fedex using legitimate logos (4.65%), Donation for diseases (1.86%) and some other random topics which includes 4.19% of overall phishing emails in our analysis.

We also analyzed another variable in our dataset such as the logos that were included in the email body. The percentage of email including logos seems significant which is more than one-fourth of the total amount (28.37%). The most frequently used logo

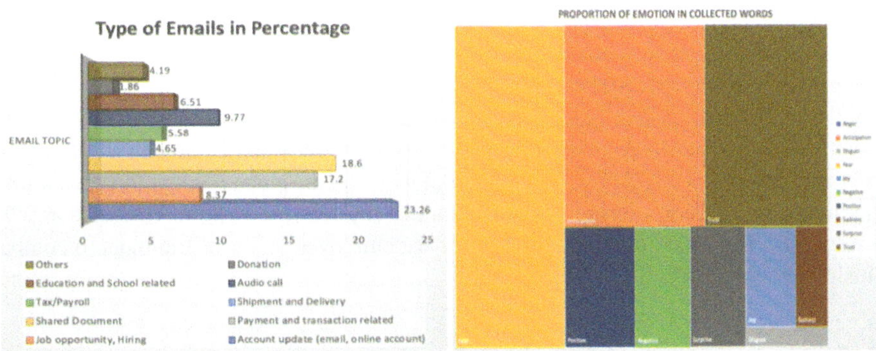

Fig. 1. Results of content analysis of phishing emails

is from Microsoft, google doc and FedEx. Another significant variable collected from archives is a URL link attached to the email body. The proportion of this occurrence is 0.637 (63.7%).

Users' Emotion Classification. To classify the emotional plea typically there are two ways of analyzing. One is semantic information that humans like to present based on what they want to say and the other is the consideration of environment and psychology [17]. For our analysis, we are following content and compelling words to extract emotional tendencies and triggers. To conduct this emotional plea extraction from context information, we input and coded each sentence as the combination of content word and emotional triggered words. For our dataset most of the sentences were content words. For emotional trigger detection, these words supply the basic emotion values or connection. In our analysis, we manually coded words used in most frequent types of phishing emails shown in Fig. 1. Then, we classify those words in the most used affect categories of the word in the lexicon: anger, anticipation, disgust, fear, joy, negative, positive, sadness, surprise, trust. Results show that fear, anticipation, and trust (29.17%, 23.61%, and 20.83%) are the most frequent emotional triggers used in our phishing email dataset

5 Discussion

In this study we conducted a systematic analysis of historical phishing emails to examine the most frequent emotional triggers and subject types used. Our results show that phishing emails often use subjects that are related to a user's online account to gain users' attention (23.26%). This type of luring is quite convincing since most users have multiple online accounts and keeping track of them is a challenge. For example, common words used in the subject of the phishing email such as, declined, suspend, confirmation, update, exceeded preys to our human limitations. Exploiting such limitations may further trigger human emotional vulnerabilities that are based on "fear", "anticipation", and "trust" [19] that we found most frequent emotion category in Fig. 1. Similarly, shared document and payment/transaction email subjects may raise curiosity for further investigation to take action which again may trigger "anticipation", "trust, and "curiosity". When we classified users' emotional triggers we find that the 3 primary exploited emotional triggers used in these emails are "fear", "anticipation" and "trust" respectively. While in this paper we report our preliminary analysis, we believe our classification of phishing email content into subject types and identification of emotional triggers often used is novel and an essential step in decreasing or preventing phishing emails. In addition, our extracted emotional triggers and frequency of words in the email content further solidify the importance of considering human factors in any security system. Furthermore, the results from this study can be used to improve employee and user trainings to combat phishing email attacks.

References

1. Krombholz, K., Hobel, H., Huber, M., Weippl, E.: Advanced social engineering at tacks. J. Inf. Secur. Appl. **22**, 113–122 (2015)
2. Ramzan, Z.: Phishing attacks and countermeasures. In: Handbook of Information and Communication Security, pp. 433–448. Springer, Heidelberg (2010)
3. Van der Merwe, A., Loock, M., Dabrowski, M.: Characteristics and responsibilities involved in a phishing attack. In: Proceedings of the 4th International Symposium on Information and Communication Technologies, pp. 249–254). Trinity College Dublin (January 2005)
4. Rekouche, K.: Early phishing. arXiv preprint arXiv:1106.4692 (2011)
5. Abawajy, J.: User preference of cyber security awareness delivery methods. Behav. Inf. Technol. **33**(3), 237–248 (2014)
6. Shackleford, D.: Cyber threat intelligence uses, successes and failures: the sans 2017 cti survey. SANS, Tech. rep. (2017)
7. Ludl, C., McAllister, S., Kirda, E., Kruegel, C.: On the effectiveness of techniques to detect phishing sites. In: International Conference on Detection of Intrusions and Malware, and Vulnerability Assessment, pp. 20–39. Springer, Heidelberg (July 2007)
8. Dong, Z., Kapadia, A., Blythe, J., Camp, L.J.: Beyond the lock icon: real-time detection of phishing websites using public key certificates. In: 2015 APWG Symposium on Electronic Crime Research (eCrime), pp. 1–12. IEEE (May 2015)
9. Sharma, T., Bambenek, J. C., Bashir, M.: Preserving privacy in cyber-physical social systems: an anonymity and access control approach (2020)
10. Rahman, S., Sharma, T., Reza, S.M., Rahman, M.M., Kaiser, M.S.: PSO-NF based vertical handoff decision for ubiquitous heterogeneous wireless network (UHWN). In: 2016 International Workshop on Computational Intelligence (IWCI), pp. 153–158. IEEE (December 2016)
11. Dong, Zhang, Y., Egelman, S., Cranor, L., Hong, J.: Phinding phish: evaluating anti-phishing tools (2007)
12. Xiang, G., Hong, J., Rose, C.P., Cranor, L.: Cantina+: A feature-rich machine learning framework for detecting phishing web sites. ACM Trans. Inf. Syst. Secur. (TISSEC) **14**(2), 21 (2011)
13. Mohammad, R.M., Thabtah, F., McCluskey, L.: Predicting phishing websites based on self-structuring neural network. Neural Comput. Appl. **25**(2), 443–458 (2014)
14. Abbasi, A., Zahedi, F.M., Zeng, D., Chen, Y., Chen, H., Nunamaker Jr., J.F.: Enhancing predictive analytics for anti-phishing by exploiting website genre information. J. Manag. Inf. Syst. **31**(4), 109–157 (2015)
15. Sharma, T., Bashir, M.: Privacy apps for smartphones: an assessment of users' preferences and limitations. In: 22nd International Conference on Human-Computer Interaction (2020)

16. Whittaker, C., Ryner, B., Nazif, M.: Large-scale automatic classification of phishing (2010)
17. Cambria, E.: Affective computing and sentiment analysis. IEEE Intell. Syst. **31**(2), 102–107 (2016)
18. Taboada, M., Brooke, J., Tofiloski, M., Voll, K., Stede, M.: Lexicon-based methods for sentiment analysis. Comput. Linguist. **37**(2), 267–307 (2011)
19. Seyeditabari, A., Zadrozny, W.: Can word embeddings help find latent emotions in text? Preliminary results. In: The Thirtieth International Flairs Conference (May 2017)
20. Shovon, A.R., Roy, S., Sharma, T., Whaiduzzaman, M.: A restful e-governance application framework for people identity verification in cloud. In: International Conference on Cloud Computing, pp. 281–294. Springer, Cham (June 2018)

Generation of User Profiles in UNIX Scripts Applying Evolutionary Neural Networks

Jairo Hidalgo[1(✉)], Cesar Guevara[2(✉)], and Marco Yandún[1(✉)]

[1] Universidad Politécnica Estatal del Carchi, Tulcán, Ecuador
{jairo.hidalgo,marco.yandun}@upec.edu.ec
[2] Universidad Tecnológica Indoamérica, Ambato, Ecuador
cesar.guevara@uti.edu.ec

Abstract. Information is the most important asset for institutions, and thus ensuring optimal levels of security for both operations and users is essential. For this research, during Shell sessions, the history of nine users (0–8) who performed tasks using the UNIX operating system for a period of two years was investigated. The main objective was to generate a classification model of usage profiles to detect anomalous behaviors in the system of each user. As an initial task, the information was preprocessed, which generates user sessions S_m^u, where u identifies the user and m the number of sessions the user has performed u. Each session S_m^u contains a script execution sequence C_n, that is $S_m^u = \{C_1, C_2, C_3,..., C_n\}$, where n is the position where the C_n command was executed. Supervised and unsupervised data mining techniques and algorithms were applied to this data set as well as voracious algorithms, such as the Greedy Stepwise algorithm, for attribute selection. Next, a Genetic Algorithm with a Neural Network model was trained to the set of sessions S_m^u to generate a unique behavior profile for each user. In this way, the anomalous or intrusive behaviors of each user were identified in a more approximate and efficient way during the execution of activities using the computer systems. The results obtained indicate an optimum pressure and an acceptable false positive rate.

Keywords: Artificial intelligence · Ehavior profiles · Neural networks · Genetic algorithm · Sistema operativo UNIX

1 Introduction

Currently, information security is one of the most important concerns in the management of information systems. Data are the most important assets of any institution and must be protected and guarded against any attempted attack or intrusion. A computer attack, as presented in [1], can be defined as an action of exploitation to take control to damage a system. The intrusion detection systems (IDS), fulfill the function of monitoring and analyzing anomalous activities on a server, on a host, or on a network.

In the work published by [3], the author proposes a cooperative cloud-based intrusion detection and prevention system called Cl-CIDPS. This system has detection, prevention,

I. Corradini et al. (Eds.): AHFE 2020, AISC 1219, pp. 56–63, 2020.
https://doi.org/10.1007/978-3-030-52581-1_8

and registration capabilities and applies signature and anomaly detection mechanisms. The Cl-CIDPS was evaluated using the network security simulator (Nessi2), which is capable of testing detection units and communication schemes.

Another article presented by [4] describes a classification method based on a type of a neural feedback network (feed-forward). This method uses an evolutionary algorithm to determine the basic structure of the coefficients of the model. The results in the testing phase applying several data sets showed that the proposed model is promising in terms of its classification accuracy and the number of coefficients of the model.

In the article published by [5], the author proposes a binomial classifier of deep learning for the network intrusion detection system. This model executes three different experiments to determine the optimal activation function, to select more important characteristics, and to test the proposed model for unseen data. The proposed classifier outperforms other models with an accuracy of 98.99% and a false alarm rate of 056% for unseen data.

This article is structured as follows. In Sect. 2, the materials and methods used to develop the proposed model are presented. In Sect. 3, the process of information analysis and the design of the behavior profile detection model are presented. In Sect. 4, the results obtained both in training and in tests with the data set are presented. Finally, the conclusions and potential future research avenues based on the research results are presented.

2 Materials and Methods

This section presents the database and the techniques used to detect user behaviors based on user profiles.

2.1 Materials

The database used was retrieved from the UCI Dataset machine-learning repository, which is comprised of nine data sets that contain the command execution history of (0–8) users who used computers with a UNIX operating system. The data set contained flags such as "** SOF **" (start of the session) and "** OF **" (end of the session) in the Shell language. The data were organized in sequential order by command and by user session. In Table 1, an example of the organization of the user records from 0 to 8 is presented. The database contains 1870 (attributes) and 11,112 sessions.

As can be seen in Table 1, the information is organized by sessions S_m^u, where u is the user who has executed the commands and m the number of sessions performed, defined as $S_m^u = \{C_1, C_2, C_3, \ldots, C_n\}$. Each S session contains n number of C_n. If the C command was executed, a value is assigned with a 1 and if it was not executed with a value of zero.

Table 1. Part of the database among users from 0 to 8 organized by commands and sessions.

usuario	sesión	Date	Date	cd	......	cmd
Usuario0	1	0	0	0	......	......
Usuario0	1	0	0	0	......	......
Usuario0	1	0	0	0	......	......
Usuario1	......	......	......	......	......	......
....	......	......	......	......	......	......
Usuario8	......	......	......	......	......	......

2.2 Methods

Genetic Algorithm with Neural Network (GANN-C)
The genetic algorithm is a global search process based on the principles of selection, crossing and mutation of elements of greater hierarchy and at least a higher generation as a result can have an improved element. These genetic algorithms belong to the Idrissi evolutionary algorithms [9, 11, 12]. It begins by mimicking human evolution and mimicking the coding of parameters, such as chromosomes, to generate new chromosomes through selection, mutation, and crossing. Throughout the process, learning is increased. The process begins by coding the chromosomes of a neural network, followed by selection, mutation, and crossing between individuals to achieve optimization and to obtain possible feasible solutions rather than a single possible solution because the selection is carried out in parallel and simultaneous searches between individuals in consideration of aptitude. It can be done using random selection, mutation, and crossing based on probabilities to obtain an optimal solution. These iterations prevent the solution from being local [10]. An accepted mathematical method is that of the minimum numbers of neurons, as shown in Eq. (1).

$$min\begin{cases} \|F(X,W,U,V)-Y\| \\ \frac{\sum_{i=1}^{N} U_i}{M_u} + \frac{\sum_{i=1}^{N}\sum_{j=1}^{ni} Vij}{M_v} \end{cases} \tag{1}$$

where N is notation, ni number of hidden layers, and X input data of the neural network. Y calculates the output of the neural network, W weights it, F is the activation function, Vij is the binary variable with a value of 1 if the neuron in a hidden layer is used, Ui is the binary variable that varies by 1 if it is hidden in the layer y 0 visible.

Greedy Stepwise Algorithm
The Greedy Stepwise algorithm is a voracious algorithm that uses a forward or backward search strategy through a subset of attribute space [13]. It can begin with all or no attributes or from an arbitrary point in space, and it stops when there is an addition or removal of any remaining attributes as a result of a decrease in the evaluation. It also generates an ordered list of attributes by traversing the space from one side to the other

and recording the order in which the attributes were selected [6]. Its generic scheme is as follows:

function voracious (C: set): set {C is the set of all candidates} S <= empty {S is the set in which the solution is built} while solution (S) and C <> empty make x <= the element of C that maximizes select (x) C <= C {x} if completable (SU {x}) then S <= SU {x} if solution (S) then return S if no return there is no solution.

Anova

The analysis of variance (ANOVA) is a statistical technique that tests the hypothesis that the means of two or more populations are equal. The authors in [7] evaluated the importance of one or more factors by comparing the means of the response variables in the different levels of the factors. The null hypothesis states that all population means (factor level means) are equal, while the alternative hypothesis states that at least one is different, as shown in Eq. (2).

$$F = \frac{\delta_1^2}{\sigma_2^2} = \frac{n\delta_{\bar{y}}^2}{\underline{S_{\bar{j}}^2}} \tag{2}$$

To test the hypothesis of the equality of the means, a statistic called F is obtained, which reflects the degree of similarity between the means being compared. The numerator of the F statistic is an estimate of the population variance based on the variability between the means of each group: $\delta_1^2 = n\delta_{\bar{y}}^2$. The denominator of the F statistic is also an estimate of the population variance but is based on the existing variability within each group: $\delta_2^2 = \bar{s}_j^2$.

Chi-Squared Test

The chi-squared test is a non-parametric test of comparison of proportions for two and more than two independent samples. It is used to determine whether there is a statistically significant difference, meaning a difference that is clearly not solely due to accidental fluctuations between the expected frequencies and the observed frequencies in one or more categories, allowing for determining whether there is a relationship between two categorical variables, as shown in Eq. (3).

$$x^2 = \sum \frac{(O - E)^2}{E} \tag{3}$$

O = refers to the observed frequencies
E = expected frequency

3 Model for Classification of Academic Information

This section describes the techniques used for data preprocessing and the description of the algorithm of the genetic neural networks for the generation of the user profile model.

3.1 Preprocessing

In this phase, data processing tasks, were performed, such as noise elimination, repeated and inconsistent data, resulting in 5266 sessions (instances) of the nine users.

For the application of the feature selection algorithms, the WEKA tool was implemented for the use of the voracious Greedy Stepwise algorithm with the result of 20 commands (attributes): 3, 8, 14, 15, 23, 190, 217, 232, 233, 400, 405, 472, 474, 804, 955, 1103, 1107, 1130, 1156, and 1586. To verify this result, the statistical selection algorithms of the Anova and chi-squared test were applied, which provided the results presented in Table 2.

Table 2. Commands selected by the Anova and chi-squared statistical tests.

No.	Chi-Squared		Anova	
1	cmd400	816752595862805	cmd8	0.0861425858949315 1
2	cmd232	3751364117020570	cmd400	0.0388112261586204 3
3	cmd9	27012477597257900	cmd232	0.03621561082360447
4	cmd217	2039787397034830	cmd15	0.03233962030561022
5	cmd405	1995484144620790	cmd9	0.029535211462100253
6	cmd15	18719177866992100	cmd405	0.026360176859279605
7	cmd175	1782630167498920	cmd217	0.024294385798907636
8	cmd182	17233192956716000	cmd1275	0.022601730526300545
9	cmd41	15399489357605700	cmd41	0.018846930231084857
10	cmd404	14620738038269700	cmd3	0.01786608618003538
11	cmd1275	1381089258698940	cmd182	0.016492048991433972
12	cmd29	13338484039656800	cmd175	0.01622306148051289
13	cmd8	13331796280961600	cmd1103	0.014946430247929743
14	cmd39	1108955493119640	cmd36	0.013282795621305077
15	cmd65	1090565048543680	cmd1136	0.011952649676878702
16	cmd3	10303072821637400	cmd1107	0.011618107721280246
17	cmd441	997528	cmd472	0.011344177212145867
18	cmd1103	9177219261406060	cmd1143	0.010707376802582269
19	cmd4	9124253695098500	cmd4	0.008961918756877574
20	cmd1107	8947306176084100	cmd1112	0.008891377860818062

3.2 Model Development with GANN-C

A genetic neural network with a configuration was applierd, as presented in Table 3.

Table 3. GANN-C configuration parameters in the intruder detection model.

Parameters	Values
Hidden layers	2
Hidden nodes	150
Eta	0.15
Alfa	0.1
Cycles	10000
Improve	0,01
Individuals	100
W range	5
Connectivity	0,5
P_bp	0,25
P_param	0,1
P_struct	0,1
Generations	100

With this configuration of the parameters of the genetic neural network and a percentage of 60% of the data set for training and 40% of the user information for the tests, highly satisfactory results were obtained in the detection of user profiles.

4 Results

A result of 80.92% (4162 instances) of correctly classified and 19.08% (1005 instances) of incorrectly classified data were obtained along with the following results obtained in the precision and the average errors as shown in Table 4.

Table 4. Results of the application of GANN-C to the data set.

Class	TP Rate	FP Rate	Precision	Recall	F-Measure	MCC	ROC Area	PRC Area
USER0	0,445	0,021	0,575	0,445	0,502	0,479	0,865	0,424
USER1	0,815	0,002	0,964	0,815	0,883	0,879	0,964	0,855
USER2	0.939	0,008	0,918	0,939	0,928	0,921	0,978	0,880
USER3	0,801	0,004	0,938	0,801	0,864	0,858	0,958	0,836

(*continued*)

Table 4. (*continued*)

Class	TP Rate	FP Rate	Precision	Recall	F-Measure	MCC	ROC Area	PRC Area
USER4	0,851	0,016	0,838	0,851	0,845	0,830	0,955	0,835
USER5	0,515	0,010	0,746	0,515	0,610	0,602	0,879	0,573
USER6	0,885	0,106	0,773	0,885	0,825	0,751	0,945	0,858
USER7	0,716	0,029	0,766	0,716	0,740	0,708	0,943	0,825
USER8	0,876	0,040	0,816	0,876	0,845	0,813	0,966	0,895
Average	**0,809**	**0,045**	**0,809**	**0,809**	**0,805**	**0,770**	**0,946**	**0,818**

5 Conclusions and Future Works

In this investigation, a model for the selection of user profiles was validated using the algorithm of genetic neural networks, which proved to be highly efficient with optimal results in precision and false positive rates.

Based on the results obtained, lines of future work are proposed. To the algorithm developed with genetic neural networks, the identification of anomalous sub-sequences can be improved by applying the algorithm based on instances, such as using Instance Selection Ranking (ISR), to improve the accuracy of the classifier as well as the detection time.

References

1. Swiler, L.P., Phillips, C., Ellis, D., Chakerian, S.: Computer-attack graph generation tool. In: Proceedings - DARPA Information Survivability Conference and Exposition II, DISCEX 2001, vol. 2, pp. 307–321 (2001)
2. Farnaaz, N., Jabbar, M.A.: Random forest modeling for network intrusion detection system. Procedia Comput. Sci. **89**, 213–217 (2016)
3. Al-Mousa, Z., Nasir, Q.: cl-CIDPS: a cloud computing based cooperative intrusion detection and prevention system framework. Commun. Comput. Inf. Sci. **523**, 181–194 (2015)
4. Martínez-Estudillo, F.J., Hervás-Martínez, C., Gutiérrez, P.A., Martínez-Estudillo, A.C.: Evolutionary product-unit neural networks classifiers. Neurocomputing **72**(1–3), 548–561 (2008)
5. Al-Zewairi, M., Almajali, S., Awajan, A.: Experimental evaluation of a multi-layer feed-forward artificial neural network classifier for network intrusion detection system. In: Proceedings - 2017 International Conference on New Trends in Computing Sciences, ICTCS 2017 (2018)
6. Disseny D'algorismes,A.I., Teresa, M., Soriano, A.: Algoritmos Voraces
7. Sow, M.T.: Using ANOVA to examine the relationship between safety & security and human development. J. Int. Bus. Econ. **2**(4), 2194–2374 (2014)
8. Lunden, O., Backstrom, M.: Stirrer efficiency in FOA reverberation chambers. Evaluation of correlation coefficients and chi-squared tests. In: IEEE International Symposium on Electromagnetic Compatibility, vol. 1, pp. 11–16 (2000)

9. Yang, J., Zhao, H., Chen, X.: Genetic algorithm optimized training for neural network spectrum prediction. In: 2016 2nd IEEE International Conference on Computer and Communications (ICCC), pp. 2949–2954 (2016)
10. Idrissi, M., Ramchoun, H., Ghanou, Y., Ettaouil, M.: Genetic algorithm for neural network architecture optimization. In: 2016 3rd International Conference on Logistics Operations Management (GOL), pp. 1–4 (2016)
11. Jenab, A,, Sari Sarraf, I., Green, D., Rahmaan, T., Worswick, M.: The use of genetic algorithm and neural network to predict rate-dependent tensile flow behaviour of AA5182-O sheets. Mater. Des. **94**, 262–273 (2016). https://doi.org/10.1016/j.matdes.2016.01.038
12. Noersasongko, E., Julfia, F.T., Syukur, A., Purwanto, P., Pramunendar, R.A., Supriyanto, C.: A Tourism arrival forecasting using a genetic algorithm based neural network. Indian J. Sci. Technol. **9**(4), 1–5 (2016)
13. Nagrikar, A., Nandagawali, S.,. Bhandarkar, A, Patle, N., Singnapure, S.: Three Way Dynamic Pricing Technique for e-Commerce Website, **2**, 31, (2018)

Use Mouse Ballistic Movement for User Authentication Based on Hilbert-Huang Transform

YiGong Zhang(✉), ShiQuan Xiong, JiaJia Li, and ShuPing Yi

Department of Industrial Engineering, Chongqing University, Chongqing 400044, China
{20142295,lijiajia,ysp}@cqu.edu.cn, xiongshquan@163.com

Abstract. In order to explore the frequency domain characteristics of mouse operation for user authentication. This paper collected experimental data on mouse ballistic movements of 10 participants on the AML website. Hilbert-Huang transform was used to extract the frequency-domain information of 9 features such as speed and acceleration during mouse movement, and formed a frequency-domain feature matrix. The Bagged-tree algorithm was used to build an authentication model. The method proposed in this paper obtained Precision = 90.25%, Recall = 88.20%. The results show that there are differences in the frequency domain information when different users operate the mouse to complete the same task, which can be used for user authentication.

Keywords: User authentication · Mouse behavior · Hilbert-huang transform · Frequency domain · Bagged trees

1 Introduction

People usually protect their accounts by setting a password for their online account. However, the incident of account property loss caused by the theft of account passwords shows that passwords are difficult to bring sufficient protection to accounts. For this reason, researchers had begun to continuously authenticate the identity of network users based on the characteristics of mouse operations of account users in time domain [1–6]. Mondal et al. [3] used mouse operation types (movement, silence, keystroke, drag), travel distance, time, and direction as the characteristics to establish a trust model and obtained a recognition accuracy of 94%. B. Wang et al. [5] established a user authentication model using random forest, and designed experiments to explore the effect of emotion on mouse operation behavior. These results showed that the mouse operation behavior of different network users was unique.

However, related researches were mainly conducted in the time domain, and there was no exploration in frequency domain. The mouse operation data can also be regarded as a signal and extended to the frequency domain for analysis. In the study of keystroke dynamics, O. Alpar [7] studied the distribution characteristics of finger tapping rhythm when different users completed the same task of typing a password, and used Short-time

I. Corradini et al. (Eds.): AHFE 2020, AISC 1219, pp. 64–71, 2020.
https://doi.org/10.1007/978-3-030-52581-1_9

Fourier Transform to analyze the frequency domain information and the differences in the frequency spectrum. A Gauss-Newton neural network was trained to test and get 4.1% EER. This study showed that even if users perform the same keystroke task, the difference in how they complete it can be detected in the frequency domain. In another study, Noy[8] observed that different people show different average jitter frequencies when they complete the same hand tracking task. Therefore, this paper explored the frequency domain characteristics of different users' mouse movements, established a authentication model, and authenticate user identity.

In this paper, an authentication method based on frequency domain information of mouse operation behavior was proposed. 10 participants were in the mouse operation experiment. Then, the Hilbert-Huang transform (HHT) was used to extract the frequency domain features of mouse operation behavior, and the Bagged-trees algorithm was used to establish a user authenticate model.

2 Method

An experiment using a mouse to perform ballistic movement was designed to investigate the frequency domain characteristics of user's mouse operation behaviors for user authentication.

2.1 Experimental Design

As shown in Fig. 1, eight number balls were evenly arranged in a circle on the experimental interface. Participants were required to operate the mouse for ballistic movements and click on the eight number balls from 1 to 8 in turn. There were ten groups of such tasks, with 20 s to relax between each group. Participants were required to use their dominant hand to operate the mouse, click each number ball accurately, cannot operate the wheel to scroll the page, and keep the mouse stable when the keystrokes were pressed. The experiment was conducted in a quiet room.

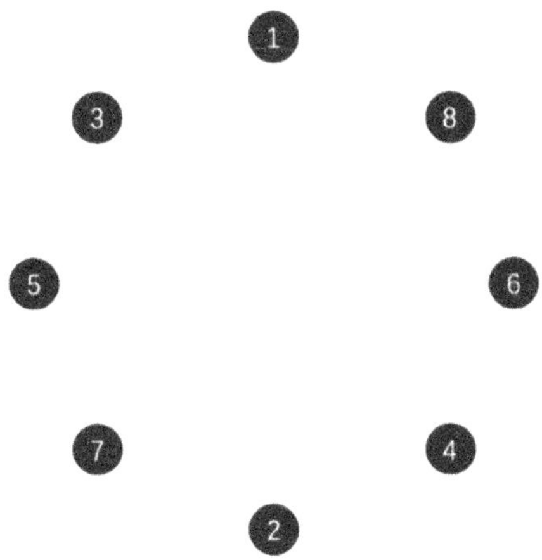

Fig. 1. Eight number balls are evenly arranged in a circle on the experimental interface

Ten naive participants (6 males and 4 females, average age 23.4, SD 1.07) were in the experiment. They were all students of Chongqing University, with more than four years of computer experience. All were right-hand, had normal or corrected-to-normal vision. All participants received adequate explanations of the merits and demerits of participating in this research and gave their informed consent.

2.2 Data Processing

Mouse data acquisition script developed using JavaScript was embedded in the website (https://www.cquieaml.com). The mouse operation data of the subjects during the experimental tasks on the website were recorded. The contents of the original data include ① Mouse operation type, ② Timestamp, ③ X-axis coordinate, ④ Y-axis coordinate, ⑤ Username. The experimental equipment was a MacBook Air laptop (macOS Mojave, 1440 × 900) and a Logitech (M330) mouse. The data analysis tools were Python and Matlab. The sampling frequency is 60 Hz.

Python was used to clean raw data. Use the ballistic movement segment between two clicks as a sample. For each sample, the horizontal velocity, vertical velocity, velocity, accelerate, jerk, snap, drop, angle of movement, angle change rate, curvature, curvature change rate were calculated. As shown in Table 1.

Table 1. Mouse movement features.

Features	Formal definition
Horizontal velocity	$v_x = \delta x/\delta t$
Vertical velocity	$v_y = \delta y/\delta t$
Velocity	$v = \sqrt{v_x^2 + v_y^2}$
Acceleration	$a = \delta v/\delta t$
Jerk	$j = \delta a/\delta t$
Snap	$s = \delta j/\delta t$
Drop	$d = \delta s/\delta t$
Angle of movement	$\theta = \arctan(\delta y/\delta x)$
Angle change rate	$acr = \delta\theta/\delta t$
Curvature	$c = \delta\theta/\delta s$
Curvature change rate	$ccr = \delta c/\delta t$

Among the 11 features, 9 time-dependent vector were used as input signals to explore their frequency domain information. In this paper, HHT was used to perform the time-frequency transformation on these input signals. HHT performs empirical mode decomposition (EMD) on the signal to obtain the intrinsic mode function (IMF), and then Hilbert Transform (HT) was used to extract the instantaneous characteristics in frequency domain of the IMFs[9].

Intrinsic Mode Function (IMF). An IMF must meet two conditions: ① for a signal, the number of extreme points and zero crossings must be equal or differ by at most one point. ② at any point, the average value of the upper and lower envelopes consisting of the local maximum and local minimum is 0.

Empirical Mode Decomposition (EMD). For the input signal $x(t)$ the EMD is:

① First find all the extreme points on $x(t)$ and connect all the maximum and minimum points with a cubic spline curve to get the upper and lower envelopes of $x(t)$ The $h_1(t)$ s obtained by subtracting the average $m_1(t)m_1(t)$ the upper and lower envelopes from the original signal:

$$x(t) - m_1(t) = h_1(t). \tag{1}$$

Repeat ① with $h_1(t)$ as the input signal until the two conditions of the IMF are met, then it becomes the true IMF component of the original input signal. Let $h_1(t) = c_1(t)$.

② The $c_1(t)$ is separated from $x(t)$ to obtain $r_1(t)$:

$$x(t) - c_1(t) = r_1(t). \tag{2}$$

The $r_1(t)$ is continued as an input signal to obtain the IMFs component until the trend component $r_n(t)$ is monotonic or has only one extreme value. At this point, the input signal $x(t)$ is decomposed into the sum of the IMFs components and a Residual. And $c_1(t)$,$c_2(t)$ … $c_n(t)$ are the IMFs component, which contains the components of different frequency bands from high to low. The process of EMD on a user's speed signal is shown in the Fig. 2.

Among these IMFs, the first IMF is usually a high-frequency component, and the Residual is the remainder after EMD that is almost a direct-current component (DC). In this study, the first IMF was discarded, which represented high-frequency signals unrelated to hand movement, and the Residual, which represented low-frequency components related to task trends. HT is then used to calculate the instantaneous frequency and instantaneous energy distribution of the remaining IMFs to obtain the instantaneous characteristics of the main IMF components of the original signal $x(t)$.

The HT defines the instantaneous frequency as:

Let $x(t)$ be a real signal, and its analytical signal is composed of Hilbert transform:

$$z(t) = x(t) + j\hat{x}(t) = a(t)e^{j\theta(t)} \tag{3}$$

Where, $\hat{x}$ is the Hilbert transform of $x(t)$, $z(t)$ is the analytical signal of $x(t)$.

$$a(t) = \sqrt{u^2(t) + v^2(t)} = |x(t)|. \tag{4}$$

$$\theta(t) = arctan\left[\frac{v(t)}{u(t)}\right]. \tag{5}$$

Instantaneous frequency f_i is defined as:

$$f_i = \frac{1}{2\pi} \cdot \frac{d\theta}{dt}. \tag{6}$$

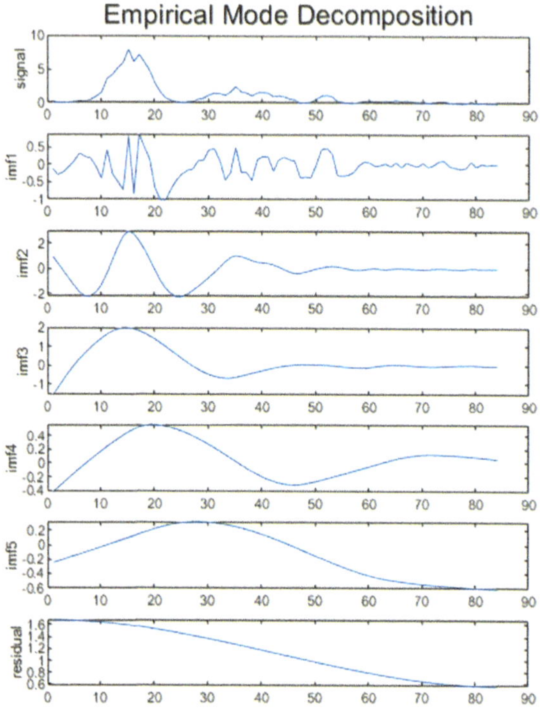

Fig. 2. The process of EMD on a user's speed signal. The first curve is the original velocity signal, the second to sixth are IMFs components, and the last is the Residual.

Which is, the instantaneous frequency of the real signal $x(t)$ is defined as the derivative of the phase of the corresponding analytical signal $z(t)$.

Instantaneous energy distribution of the IMF $c_i(t)$ is defined as:

$$E_i(t) = \frac{1}{2} \cdot a_i^2(t). \tag{7}$$

$$E(t) = \sum E_i(t). \tag{8}$$

Where, the $E(t)$ represents the instantaneous energy value of the signal at any time t. It describes the energy transfer and fluctuation of the signal at different times.

Then, each linear motion segment is taken as the object, and the maximum value, minimum value, average value, standard deviation, and range of the instantaneous frequency and instantaneous energy are calculated. Each motion segment will produce $9 * 2 * 5 = 90$ features, the feature matrix shown in Table 2.

Table 2. Feature Pool in Frequency Domain

Feature origin		Transient features in Time-Frequency domain		
Horizontal velocity				
Vertical velocity				
Velocity				Minimum
Acceleration		Instantaneous frequency		Maximum
Jerk	×		×	Mean
Snap		Instantaneous energy		Standard deviation
Drop				Range
Angle of movement				
Angle change rate				

3 Results

The experimental data was divided into a training data set and a test data set in a 4: 1 ratio. Bagged-tree algorithm (200 trees) was used to develop a trusted model of mouse behavior and verify the user's mouse operation data. The accuracy of the model was 89.30%. Precision = 90.25%, Recall = 88.20% (Table 3).

$$Precision = \frac{TP}{TP + FP}. \quad (9)$$

$$Recall = \frac{TP}{TP + FN}. \quad (10)$$

Table 3. Confusion matrix

User	A	B	C	D	E	F	G	H	I	J	Precision	Recall
A	16		5					1			94.12%	72.73%
B		25									100.00%	100.00%
C	1		21					1			80.77%	91.30%
D				20							100.00%	100.00%
E					23			2	3		79.31%	82.14%
F						25				1	96.15%	96.15%
G							31				91.18%	100.00%
H								28	1	1	96.67%	93.33%
I					6		3	1	10		71.43%	50.00%
J						1				26	92.86%	96.30%
Average											90.25%	88.20%

The TP is True positive (number of positive classes predicted as positive).
The FP is False positive (number of negative classes predicted as positive).
The FN is False negative (number of positive classes predicted to be negative).

4 Discussion and Conclusion

In this article, a user authentication method based on HHT was proposed. The mouse operation data came from the ballistic movement by subject's operation with the mouse. Then, the HHT was used to extract the frequency domain information of time domain features such as speed, acceleration, and angular velocity. Finally, the Bagged-tree algorithm was used to establish the authentication model.

The results of this research showed that mouse operation behavior was also difference in the frequency domain, and the method proposed in this paper could effectively complete user authentication. At the same time, this study provided a new way for the study of mouse dynamics and expanded the feature space of mouse behavior.

In the future, we hope to further improve the effectiveness of this method. Features in the time domain will be added to form a matrix of mouse behavior features in the time-frequency domains. On the other hand, the data in the experimental environment were only the mouse behavior under a single login, they were difficult to characterize the mouse operation behavior habits of users who had used their accounts for long-term trustworthy interaction[10]. For this reason, this research will be extended to non-experimental environments to explore the long-term legal login habits of users to further protect safety of network accounts.

Acknowledgments. This work was supported by the National Natural Science Foundation of China under Grant No. 71671020.

References

1. Zheng, N., Paloski, A., Wang, H.: An efficient user verification system using angle-based mouse movement biometrics. ACM Trans. Inf. Syst. Secur. **18**(3), 1–27 (2016). https://doi.org/10.1145/2893185
2. Feher, C., Elovici, Y., Moskovitch, R., Rokach, L., Schclar, A.: User identity verification via mouse dynamics. Inf. Sci. **201**, 19–36 (2012). https://doi.org/10.1016/j.ins.2012.02.066
3. Mondal, S., Bours, P.: A computational approach to the continuous authentication biometric system. Inf. Sci. **304**, 28–53 (2015). https://doi.org/10.1016/j.ins.2014.12.045
4. Bailey, K.O., Okolica, J.S., Peterson, G.L.: User identification and authentication using multi-modal behavioral biometrics. Comput. Secur. **43**, 77–89 (2014). https://doi.org/10.1016/j.cose.2014.03.005
5. Wang, B., Xiong, S., Yi, S., Yi, Q., Yan, F.: Measuring network user trust via mouse behavior characteristics under different emotions, In: HCI for Cybersecurity, Privacy and Trust, pp. 471–481. Cham (2019)
6. Salman, O.A., Hameed, S.M.: Using mouse dynamics for continuous user authentication. In: Arai, K., Bhatia, R., Kapoor, S. (eds.) Proceedings of the Future Technologies Conference (FTC) 2018, vol. 880, pp. 776–787. Springer, Cham (2019)

7. Alpar, O.: Frequency spectrograms for biometric keystroke authentication using neural network based classifier. Knowl.-Based Syst. **116**, 163–171 (2017). https://doi.org/10.1016/j.knosys.2016.11.006
8. Noy, L., Alon, U., Friedman, J.: Corrective jitter motion shows similar individual frequencies for the arm and the finger. Exp. Brain Res. **233**(4), 1307–1320 (2015)
9. Huang, N.E., et al.: The empirical mode decomposition and the Hilbert spectrum for nonlinear and non-stationary time series analysis. In: Proceedings of the Royal Society of London. Series A: Mathematical, Physical and Engineering Sciences, vol. 454, no. 1971, pp. 903–995, March 1998, https://doi.org/10.1098/rspa.1998.0193
10. Zhang, L.: Research on agent-based human-information system trusted interaction in distributed cooperative work environment. TOAUTOCJ **3**(1), 1–7 (2011). https://doi.org/10.2174/1874444301103010001

Awareness, Training and Education

Understanding and Enabling Tactical Situational Awareness in a Security Operations Center

Ryan Mullins(✉), Ben Nargi, and Adam Fouse

Aptima Inc., 12 Gill Street Suite 1400, Woburn, MA 01801, USA
{rmullins,bnargi,afouse}@aptima.com

Abstract. Cybersecurity operations are highly complex, requiring the coordination of specialized skills across multiple teams to successfully execute missions. Command and control within security operations centers is dominated by fragile mental models, demonstrating a need for systems that reinforce shared situational awareness across the organization. In this paper, we present the results of our research to: (1) define the needs associated with tactical cyber situational awareness; and (2) evaluate the usability and utility of a prototype tactical situational awareness dashboard. We found that incident tracking, tasking structure, execution timeline, and resource health constitute the essential aspects of tactical cyber situational awareness. Evaluations of prototypes suggest that three visualizations are well suited for conveying this information. We believe these results generalizable and will enable the development of tactical situational awareness capabilities in Security Operations Centers across public and private enterprises.

Keywords: Cybersecurity · Situational awareness · Visual analytics · Command and control

1 Introduction

Situational awareness in the cybersecurity domain is often framed as an operator's perception of their environment and events within it given the specific context of defender-attacker interactions [1–4]. However, this framing captures a relatively narrow slice of the Defensive Cyber Operations (DCO) landscape, with a very externally-focused viewpoint. Preventative maintenance, research and development, and vulnerability analysis are just some of the additional, internally-focused functions that make real-time cyber defense possible and increase the scope of needed situational awareness.

Our research seeks to examine cyber situational awareness from a different vantage point, one concerned with enabling tactical coordination and collaboration between these functions as executed within a unified organization. From this perspective, shared understanding [5, 6] becomes the critical factor in enabling situational awareness, as multiple operators with varying needs will be using common tools to build their understanding and awareness of how the actions of their function can impact or be impacted

I. Corradini et al. (Eds.): AHFE 2020, AISC 1219, pp. 75–82, 2020.
https://doi.org/10.1007/978-3-030-52581-1_10

by the actions of others. We are particularly interested in the situational awareness needs of those working in a Security Operations Center (SOC). SOCs have emerged as an industry best practice for centralizing command and control (C2) and execution of an organization's response to incidents and events [7].

In this paper, we present the results of research to understand the tactical situational awareness needs of a DCO unit within the United States Air Force. Our research objectives were two-fold. First, we sought to develop a consensus definition of, and requirements for, tactical situational awareness among C2 personnel within the unit. Second, we sought to evaluate the usability and utility of a prototyped system enabling tactical situational awareness.

2 Defining Tactical Situational Awareness in Cybersecurity

2.1 Methods

We used a combination of structured and semi-structured interview methods in individual and focus group formats over three engagements. The first engagement deconstructed the unit's Planning, Briefing, Execution, and Debriefing (PBED) process, which defines the information, reporting structure, and communications flows. The second engagement involved shadowing and interviewing Crew Commanders through each phase in the PBED process. The final engagement was a focus group to refine and expand upon the findings from the prior two engagements.

Participants varied throughout the course of the research. The population sizes for the three engagements were three, one, and five, respectively. While these numbers are relatively small, the third engagement included a sufficiently representative sample of to validate the definitions and requirements given unit and organizational norms.

2.2 Results

Interview transcripts were synthesized to develop the final definition of tactical cyber situational awareness: ***the perception of the missions, tasking, resources, status, interactions, and correlates between constituent teams executing DCO***. Here, we describe the constructs that organize these situational awareness needs of the unit. We have identified two use cases for tactical situational awareness visualizations:

1. The ***Operations Floor Dashboard***, where the visualization is presented on a large-format display at the front of the operations floor for all operators; and
2. The ***Leadership Dashboard***, where the visualization is presented on desktop workstations used by commanders and mission leads.

Categorized Incidents and Events. The US Government differentiates between two types of violations of cybersecurity policies and practices, incidents and events [8], and categorizes and prioritizes them based on the nature of the violation [9]. Categorized incidents and events (CAT Events) constitute the bottom-up tasking for DCO units. All tactical cyber situational awareness tools must be able to visually represent CAT Events.

The critical data include the CAT Event's priority, primary and secondary incident categories, the responsible person, and the action being taken. Due to the collaborative nature of response, CAT Events often span multiple execution windows. Participants recommended visualizing CAT Events in a tabular format for all use cases.

Tasking. Tasking refers to the work included in a unit's mission plan, and represents the top-down unit of work for DCO units. Situational awareness tools must be able to represent the essential elements of the plan, as well as the status, responsible person, and action being taken. The essential elements, shown in Fig. 1, include the packages, missions, and tactical tasks included in the plan. Packages and missions are associated with discrete objectives that describe actions required to achieve the desired end state. Each objective is associated with measures of effectiveness (MOEs), used to assess their status (e.g., achieved, partially achieved, failed). Tactical tasks are associated with measures of performance (MOPs), used to assess their status (e.g., complete, failed, in progress, not started). For both use cases, we recommend itinerary-style visualizations that contextualize MOEs and MOPs within a mission. Missions provide a meaningful anchor to visualize subordinate tactical tasks, or to link out to packages being executed in tandem with partners. Participants recommend limiting the visibility of MOEs and MOPs to the Leadership Dashboard, as it would reduce operator stress on the floor.

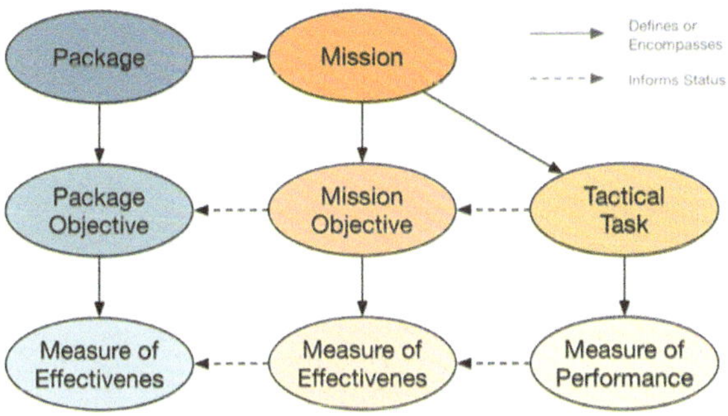

Fig. 1. Discrete elements of a DCO mission plan that must be tracked by tactical situational awareness systems. All links between nodes represent one-to-many relationships.

Execution Timeline. The execution timeline represents the temporal context of operations, emphasizing the structure and sequencing of missions and their subordinate tasks. Participants strongly recommended that a Gantt chart be used for the timeline in both use cases. The vertical axis of the chart should be divided using a categorical hierarchy where the major categories are the missions and the minor categories are the tactical tasks. The label should also include the name of the lead for each tactical task. Authorized service interruptions should be denoted on the chart to provide additional context regarding the rationale behind task sequencing.

Resource Health. Resource health relates to status of the individual, tangible components associated with operations, including hard, soft, and human resources. Considering the complexities of DCO, the list has the potential to reach unmanageable proportions. Therefore, we engaged participants in the task of scoping the essential elements to visualize: *missions*, *tasks*, and *cyber key terrain*. Health measures for missions and tasks are the aggregate status of the associated MOEs and MOPs. Health measures for cyber key terrain vary based on component type and sensor characteristics. Participants recommended that health measures be conveyed using a traffic light protocol (i.e., green = good, yellow = warning, and red = dangerous). Participants prefer visualizations that include status and trend simultaneously. The base visualization should be the same for both use cases, but the Leadership Dashboard should provide detailed exploration features, aligning with visual analytics best practices [10].

3 Evaluating a Prototype Situational Awareness Capability

3.1 Concept and Design

This system, shown in Fig. 2, was designed to support three needs: tasking, resource health, and CAT Event tracking.

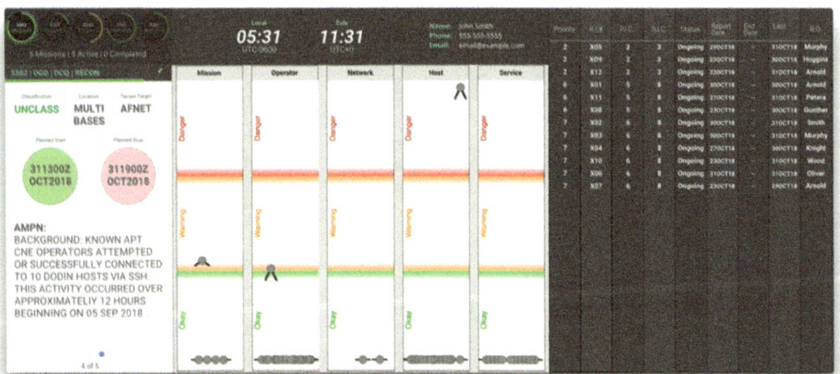

Fig. 2. The prototype tactical cyber situational awareness prototype as implemented and evaluated, showing the tasking carousel (left), *Aquarium* (center), and CAT Event tracker (right).

Tasking. Tasking structure is represented by the ***Tasking Carousel*** visualization (Fig. 2, left). Each mission in the plan is represented as a circular indicator at the top of the carousel, which displays the mission's identifier and classification. The stroke of the indicator conveys the status of that mission (e.g., green, yellow, or red) and the arc length of which conveys mission progress. Below the carousel is a display of how many missions are included in the current plan, broken down by active and completed. Below this is the detailed information for the indicator in the first position of the carousel: the execution window, location, terrain, tactical tasks, and background information.

Resource Health. Resource health is represented by a novel visualization technique we call the ***Aquarium*** (Fig. 2, center). The Aquarium represents discrete operational resources as *Fish* that are binned into *Tanks*. The prototype tanks were divided by resource type, to include missions, operators, networks, hosts, and services. Tanks are divided into the three sections of a traffic light protocol. The analogy is that healthy fish live in the green area near the bottom and unhealthy fish live in the red at the top. The horizontal position of a fish is not meaningful and is determined using the Halton sequence [11] to minimize obfuscation. Each fish has two *fins* used to indicate trajectory. The length and angle between the fins is used to indicate the rate of travel using five discrete states to optimize scan-ability [12]: (1) long-close-below for rapidly unhealthy; (2) medium-spread-below for slowly unhealthy; (3) short-wide-center for none; (4) medium-spread-above for slowly healthy; and (5) long-close-above for rapidly healthy.

Categorized Incident and Event Tracking. CAT Events are represented in a table visualization (Fig. 2, right). Each row represents a CAT event with columns for priority, report identification number, primary and secondary incident categories, status, report date, end date, date of last action, and responsible operator. CAT Events are sorted in priority order (ascending order of primary category, see the scale definitions in [9]).

3.2 Methods

The situational awareness prototype was designed at a high level of fidelity meant to simulate an actual software experience. The content and data were generated and validated by subject matter experts familiar with USAF DCO operations and the specific context of the participating unit. The prototype was evaluated using an online survey that included static and dynamic stimuli, and divided into sections focused on the validation of design assumptions, the usability of the visualizations in isolation, and the utility of the visualizations in concert. Following the survey, we engaged participants in a final group discussion to gather qualitative feedback about the design and functions.

Participants included four crew commanders with 1-6 years of experience in the role. As with the knowledge elicitation sessions, we recognize that the small sample size will inhibit the meaningful statistical assessment of these results, but we believe it sufficient to validate the concepts and functions of the prototype [13].

3.3 Results

We describe specific feedback below, but we would like to highlight two general lessons. First, due to the diversity of operator responsibilities, it is preferred that experiences for the *Operations Floor Dashboard* use case are minimally adaptive. This provides a more consistent frame of reference that participants believe will be more resilient [14] given environmental dynamics. Second, proper tactical situational awareness cannot be achieved without supporting both the *Operations Floor Dashboard* and the *Leadership Dashboard* use cases, the latter of which should be more adaptive.

Design Assumptions. Questions were designed to assess how useful information is (Table 1) and how frequently information would be used (Table 2) from the perspective of a crew commander and an operator.

Usefulness. The crew commander perspective aligned well with the focus group. The most useful information are the missions and their status, the mission leads, and the CAT Events and their status. The operator perspective diverged slightly, with the most useful information being the health and status of cyber key terrain, the missions and their status, and the missions leads.

Table 1. Most useful information (top 3).

Rank	Commander	Operator
1	Missions and status	Cyber key terrain
2	Mission leads	Missions and status
3	CAT events and status	Mission leads

Frequency. The top three for crew commanders are mission status, CAT Event status, and missions being executed, and the top three for operators are the CAT Event status, the hard resource status, and the mission status.

Table 2. Most frequently used information (top 3).

Rank	Commander	Operator
1	Mission status	CAT events and status
2	CAT events and status	Had resource status
3	Mission list	Mission status

Content and Representations. The next three sections dealt with the usability of the visualizations in isolation. Generally, the results were positive, but designs could be improved by adding information about tactical tasks and sequencing. This finding was particularly true of the ***Tasking Carousel*** visualization, which participants noted could be simplified to a structured checklist that included tactical tasks, task leads, and status.

The CAT Event table was considered usable, but operators had some disagreement as to the importance (order) and inclusion of some fields. Priority, incident categories, report identifier, responsible operator, and next action are essential. The utility of timing information, such as report date, last update, and end date, was debated. Further exploration of these fields will be conducted in future research.

The usability and utility assessment for the ***Aquarium*** visualization was divided into two components, the *Fish* in isolation and the specific *Tanks* configuration. *Fish* visualizations were assessed for comprehension of the direction and rate of travel. All respondents were able to correctly infer both attributes. Participant feedback on the *Tank* content and configuration was mixed, with most feeling that Missions, which were grouped into a single *Tank*, should each receive their own *Tank* with *Fish* representing Tactical Tasks. Participants felt that cyber key terrain was represented appropriately, but noted that these should be configurable based on the nature of operations.

4 Discussion

Diversity of activity within the SOC requires a more introspective focus than is provided by other situational awareness tools [1–4]. The central need for tactical situational awareness is to understand coordination and collaboration, requiring the correct representation of constituent tasking- and resource-centric data, especially regarding their status and sequencing. Representations of tasking-centric elements must account for top-down and bottom-up directives. Representations of resource-centric elements are less critical, and should play a diminished role in interface design. Finally, there is a meaningful distinction between the situational awareness needs of commanders and operators. System designs should provide relatively static experiences for shared displays and relatively dynamic experiences for individualized displays.

Finally, we found it important to emphasize the power of redundancy. This was most present in our representations of tasking-centric information. This approach is similar to linked, coordinated visualization in visual analytics [15], but differs in that the *Operations Floor Dashboard* does not provide the interactivity to enable exploration.

Acknowledgments. This work was conducted in connection with contract FA8750-17-C-0203 with the Air Force Research Laboratory. The views, opinions, and findings contained herein are those of the authors alone. The authors would like to thank Rick Fedors, Capt. Jonathan Beabout, TSgt. Freddie Morales-Torres, Robert Hoffman, Larry Bunch, John Schrimpf, and Mike Garrity for their contributions to this work.

References

1. Endsley, M.R.: Design and evaluation for situation awareness enhancement. In: Proceedings of the Human Factors Society Annual Meeting, vol. 32, no. 2, pp. 97–101. SAGE Publications, Los Angeles (1988)
2. Franke, U., Brynielsson, J.: Cyber situational awareness–a systematic review of the literature. Comput. Secur. **46**, 18–31 (2014)
3. Jajodia, S., Noel, S., Kalapa, P., Albanese, M., Williams, J.: Cauldron mission-centric cyber situational awareness with defense in depth. In: MILCOM, pp. 1339–1344 (2011)
4. Matthews, E.D., Arata III, H.J., Hale, B.L.: Cyber situational awareness. Cyber Def. Rev. **1**(1), 35–46 (2016)
5. Entin, E.E., Serfaty, D.: Adaptive team coordination. Hum. Factors **41**(2), 312–325 (1999)

6. MacMillan, J., Entin, E.E., Serfaty, D.: Communication Overhead: The Hidden Cost of Team Cognition. Team Cognition: Process and Performance at the Inter- and Intra-Individual Level. American Psychological Association, Washington, DC (2004)
7. Sundaramurthy, S.C., Case, J., Truong, T., Zomlot, L., Hoffmann, M.: A tale of three security operation centers. In: Proceedings of the 2014 ACM Workshop on Security Information Workers, pp. 43–50. ACM (2014)
8. Cichonski, P., Millar, T., Grance, T., Scarfone, K.: Computer security incident handling guide. NIST Spec. Publ. **800**(61), 1–147 (2012)
9. Cyber Incident Handling Program, CJCSM 6510.01b, Joint Chiefs of Staff, Washington, D.C. (2012)
10. Shneiderman, B.: The eyes have it: a task by data type taxonomy for information visualizations. In: Proceedings of the 1996 IEEE Symposium on Visual Languages. IEEE (1996)
11. Halton, J.H.: Algorithm 247: radical-inverse quasi-random point sequence. Commun. ACM **7**(12), 701–702 (1964)
12. Bennett, K.B., Flach, J.M.: Display and Interface Design: Subtle Science. Exact Art. CRC Press, Boca Raton (2011)
13. Virzi, R.A.: Refining the test phase of usability evaluation: how many subjects is enough? Hum. Factors **34**(4), 457–471 (1992)
14. Woods, D.D.: Essential characteristics of resilience. In: Resilience Engineering, pp. 33–46. CRC Press (2017)
15. Roberts, J.C.: State of the art: coordinated & multiple views in exploratory visualization. In: IEEE Fifth International Conference on Coordinated and Multiple Views in Exploratory Visualization CMV 2007, pp. 61–71 (2007)

Cybersecurity Risks and Situation Awareness: Audit Committees' Appraisal

Stéphanie Thiéry[1,3](✉) and Didier Fass[1,2]

[1] ICN BS, 86 rue du Sergent Blandan, Nancy, France
stephanie.thiery@icn-artem.com
[2] Mosel Loria, UMR CNRS 7503, Université de Lorraine, Nancy, France
didier.fass@loria.fr
[3] CEREFIGE, EA 3942, Université de Lorraine, Nancy, France

Abstract. The issue of cybersecurity has become a challenge for companies and boards of directors. Cybersecurity is not only an IT topic, but a risk extended to all operations of the companies. Indeed, cybersecurity potentially has an impact on financial reporting quality, this attribution being one of the duties of audit committees. Using Endsley's model, our exploratory study seeks to determine the levels of cyber situational awareness of audit committee members, how they comply with it and if this appraisal matches the steps identified within the model.

Keywords: Cybersecurity awareness · Board of directors · Audit committee · Safety by design · Cyber-resilience

1 Introduction

Cybersecurity is nowadays a significant topic within organizations since the last 25 years, assets of companies have evolved from physical assets to the digital [1]. Intangible assets valued according to international standards are particularly sensitive to internal or external manipulation and attack. However, taking cyber risk into account mainly covers the IT (internal IT) technical risks. The human and organizational factor aspects are neither clearly known nor clearly identified, in particular by decision-making bodies such as Board of Directors. Thinking the company as an integrated system "critical security" and the risks inherent to its field of activity are a theoretical and practical issue.

If the explicitly described missions by regulation were discussed in the literature, only few studies related to cyber-attacks management exist at board level [2] or, specifically, at audit committee level. Thus, the literature on boards and audit committees has not operationalized the examination of this risk by governance institutions. Indeed, the criticism of organizational data is a well-known issue of actors who are responsible for, as indicated by [3], since the developed model in their study allows to take into account the data owners', senior management's and legal experts' point of view to give a framework to data security assessment. Authors also recommend an implication of internal audit and information technology functions [4].

I. Corradini et al. (Eds.): AHFE 2020, AISC 1219, pp. 83–87, 2020.
https://doi.org/10.1007/978-3-030-52581-1_11

However, these works do not consider the direct appropriation of this issue by governance institutions as the board of directors or audit committees. Yet, regulation seems to have integrated the topic, requiring firms to perform a cyber-risk assessment, with associated costs and consequences or a description of occurred cybersecurity issues including their costs and consequences [5]. Clark and Harrell [6] however highlight that if SEC (Securities and Exchange Commission) current recommendations (disclosure of data breach issues) became obligations, directors of public companies could incur lawsuit risks, in addition to the decrease of share price. For this reason, Lunn [7] indicates the question directors should ask in case of cyber-attack in order to protect their responsibility if the latter was engaged. He advocates considering some factors in their monitoring role: existence of monitoring process of cyber risks, probability and consequences of loss related to cyber-attacks, existence of consequences which may adversely affect human lives or the survival of organization or implementation of action plans to mitigate cyber risks. Recommendations are given in order to limit directors' responsibilities, such as training directors to cybersecurity issues or recruiting directors having an experience in this field. Von Solms [8] takes up an assessment model of board maturity in terms of cybersecurity device review. This model allows an assessment of cyber governance knowledge within board members, giving insights on their understanding of the issue and the implementation degree of cyber governance. However, if this model contributes to the self-assessment of the directors' actions, it seems that it does not exist any study building a state of directors' competences and received information (particularly of audit committee members) in terms of cybersecurity. To our knowledge, the appraisal of audit committee members on the review of cyber risks has not been investigated in the literature: are auditing committee members aware of issues related to cybersecurity?.

There are still unanswered questions regarding the audit committee functioning process in general [9] and, specifically, as for cyber issues and those process issues remain neglected by researchers. In order to answer our previous question, we favored a field study and a qualitative approach. This leads us to determine if audit committees address significant cybersecurity topics and/or face cybersecurity breaches. Our work in progress seeks to determine if audit committee members are aware of the issues linked to cybersecurity, in order to improve both cyber-resilience and safety by design decision-making.

2 Cybersecurity Awareness and Human Factors at Audit Committee Level

Management and production information systems and digital information, especially if they are strategic assets for the company, are safety critical. Consequently, the impact of deficiencies of cybersecurity can have global consequences: loss of intellectual property, risks of legal or regulation-linked penalties, reputation loss, costs to restore clients' confidence and to give explanations to authorities [10]. To prevent that systemic risk, one needs to be able to estimate related human factors such as situation awareness and processes responsible for maintaining it.

Three categories of main components impact cybersecurity at each organizational levels of the socio-technical system: technical risks/ factors, human risks/ factors and organizational risks/ factors. Misunderstanding or underestimation of cyber-risks is thus a danger for the company that must be dramatically considered at board level.

Because cyber risks are not only virtual but actual, taking into account this serious danger is a question of situation and risk awareness depending on knowledge, cognitive bias or emotional states which participate to the perception of the risks and which influence decision-making and control processes [11–13].

Just like aeronautics, e.g. airplane piloting and air traffic control, enhancing cyber situational awareness at board and audit committee levels is a major issue. Thus, board of directors seem to be considering the topic since, according to [1], 81% of surveyed boards address cybersecurity issues during meetings and that 51% of respondents claim that cybersecurity should be considered at audit committee level. In order to evaluate the cyber situation awareness of audit committees' members, we favored Endsley's framework [14] and its three levels (level SA). This enables us to assess the perception, the comprehension and the projection of the audit committee's members.

3 Directors' Appraisal of Cyber Issues

3.1 Method Used, Data Collection and Analysis

Our collection of empirical material is driven by our exploratory study. We both rely on invaluable observations of audit committee meetings, interviews with audit committee members and participants, publicly available reports and internal documents. Interviews lasted, on average, between 60 and 120 min and on-site observations around 150 min each.

First, we reviewed publicly available documents (10-K reports) to gain understanding on how organizations formally report on audit committees' appraisal of cyber issues. We next got closer to the field and supplemented our empirical material, with a source of data constituted by on-site observations of two audit committee meetings. Furthermore, we carried out 27 semi-structured interviews with audit committees' members but also with individuals who attend the meetings, such as partners of audit firms and chief audit executives. Interviews are a relevant data collection mechanism, complementary to observation-based material [15]. Having completed on-site observations with, first, documentation and, second, interviews, were a powerful tool in order to help us gathering evidence of their appraisal of cyber risks and cyber issues.

3.2 Level 1 SA - Perception: Disclosed Cyber-Awareness to the Public and Individual Returns of Experiences

Listed firms communicate and disclose both their internal control concerns and their risk assessment. Being part of the most important emerging risks, cyber issues are disclosed within the 10-K reports. This is confirmed by the content analysis we achieved on our 2015–2016 annual reports of French firms. On 66 firms for which we examined the audit committee reports, 18 made explicitly reference to a review of cyber risks (Table 1). Out of our 2 on-site observations and 27 interviews conducted, only one on-site observation and three interviewees mentioned and analyzed cyber issues. This is far more less than the 51% of firms supposed to address cyber risks at the level of the audit committee [1]. Hence, according to our fieldwork, some audit committee members highlighted a basic perception of cyber situation awareness: "*we have an extremely high risk (…) on particular points which can be presented and studied in depth by the audit committee*".

Table 1. Content analysis (66 listed French firms).

Firms addressing cybersecurity topic/cyber risks	Firms hearing the Head of IT departments during audit committees
Arkema, Biomérieux, Bouygues, Burelle, CGG, Dassault Systèmes, Engie, Essilor, L'Oréal, Renault, Saft, Sanofi, Technicolor, Total (*14 firms*)	Endered, Essilor, Saft, Valeo (*4 firms*)

3.3 Level 2 SA - Comprehension

However, it seems that only 54% of global organizations have carried out an assessment related to fraud or economic crime. In particular, less than half of firms have achieved a vulnerability assessment related to cyberattacks and only 30% have implemented an action plan [16]. Furthermore, for members of audit committees and, more generally, for boards, "cyber" is new for many directors, and is certainly far from intuitive" [17].

As our interviewees stated: "*we have to perform regular checkups and (…) Basically, we acknowledge that some persons may have non-restricted accesses to the system (…) This must be the subject of a presentation and in-depth study while audit committee meetings*".

3.4 Level 3 SA - Projection

Mostly our field work highlights that directors are first "cyber-risks aware" and that they intend to get a specific overview of the main cyber issues, using ever specialists or governmental agencies in order to help them appraise and improve cybersecurity: "*we asked specialists and ANSSI in order not to waste time*".

Moreover, our interviewees assess that, in order to be compliant with the main internal control frameworks (COSO, COBIT), they target some specific levers of control, such as control environment (the 'tone at the top' and human knowledge and skills) and control activities (Segregation of duties): "*we need to train our people to improve their cyber awareness and secure their accesses and behaviors(…) specifically we must disseminate this cyber awareness through operational middle management and their teams*".

4 Conclusion

Annual reports should disclose the risks including cyber issues if it happened but only if they are material. This means that without any material effect, cyber issues are not always revealed to the public. Our analysis confirms that this disclosure is not obvious and depends on the knowledge, expertise and will of the boards. Nonetheless, our exploratory study highlights that, when cyber issues are tackled by audit committees, they follow Endsley's process and that they both embrace, appraise, evaluate and disseminate the issues. Our preliminary analysis should be, of course, deepened with archival data in order to validate this fieldwork evidence, but our work underscores requirements and impetus for improving board cyber situation awareness.

Acknowledgments. We thank Marion Tellechea for research support and La Région Grand Est, Pacte Grandes Ecoles, for financial support.

References

1. NACD: Cyber-Risk Oversight. In: Clinton, L. (ed.) Director's Handbook Series, National Association of Corporate Directors, Washington DC, USA (2017)
2. Higgs, J.L., Pinsker, R., Smith, T., Young, G.: The relationship between board-level technology committees and reported security breaches. J. Inf. Syst. **30**(3), 79–98 (2016)
3. Rahimian, F., Bajaj, A., Bradley, W.: Estimation of deficiency risk and prioritization of information security controls: a data-centric approach. Int. J. Account. Inf. Syst. **20**, 38–64 (2016)
4. Steinbart, P.J., Raschke, R.L., Gal, G., Dilla, W.N.: The influence of a good relationship between the internal audit and information security functions on information security outcomes". Account. Organizations Soc. **71**, 15–29 (2018)
5. CF Disclosure Guidance: Topic No. 2 - Cybersecurity - SEC.gov (2011). https://www.sec.gov/divisions/.../guidance/cfguidance-topic2.htm
6. Clark, M.E., Harrell, C.: Unlike chess, everyone must continue playing after a cyber-attack. J. Investment Compliance **14**(4), 5–12 (2013)
7. Lunn, B.: Strengthened director duties of care for cybersecurity oversight: evolving expectations of existing legal doctrine. J. Law and Cyber Warfare **4**(1), 109–137 (2014)
8. Von Solms, B.: Towards a cyber governance maturity model for boards of directors. Int. J. Bus. Cyber Secur. (IJBCS) **1**(1), 1–9 (2016)
9. Gendron, Y., Bédard, J., Gosselin, M.: Getting inside the black box: a field study of practices, «Effective» Audit Committees. Auditing: J. Pract. Theory, **23**(1), 153–171 (2004)
10. KPMG, Boardroom Questions. Cybersecurity - What does it mean for the board (2017). https://home.kpmg/content/dam/kpmg/be/pdf/boardroomquestions/boardroom-questions-cyber-security-what-does-it-mean-for-the-board.pdf
11. Endsley, M.R.: Toward a theory of situation awareness in dynamic systems. Hum. Factors J. **37**(1), 32–64. Human Factors: J. Hum. Factors Ergon. Soc. **37**, 32–64 (1995a)
12. Endsley, M.R.: Measurement of situation awareness in dynamic systems. Hum. Factors: J. Hum. Factors Ergon. Soc. **37**(1), 65–84 (1995)
13. Damasio, A.: Descartes' Error: Emotion, Reason and the Human Brain. Putnam Publishing, New York (1994)
14. Endsley, M.R.: Situation awareness analysis and measurement, chapter theoretical underpinnings of situation awareness. a critical review. In: Endsley, M.R., Garland, D.J. (eds.) Situation Awareness Analysis andMeasurement, pp. 3–33. Lawrence Erlbaum Associates, Mahwah (2000)
15. Yin, R.K.: Case Study Research Design and Methods. Sage, Thousand Oaks (2014)
16. PwC's Global Economic Crime and Fraud Survey (2018). https://www.pwc.com/gx/en/services/advisory/forensics/economic-crime-survey.html
17. Advisor, The Corporate Governance: Cybersecurity **2**, 5 (2014)

Addressing Human Factors in the Design of Cyber Hygiene Self-assessment Tools

Jacob Esparza[1](✉), Nicholas Caporusso[2], and Angela Walters[1]

[1] Department of Informatics, Fort Hays State University, 600 Park Street, Hays 67601, USA
jtesparza@mail.fhsu.edu, awalters@fhsu.edu
[2] Department of Computer Science, Northern Kentucky University, Louie B Nunn Dr, Highland Heights 41099, USA
caporusson1@nku.edu

Abstract. As cybersecurity (CS) threats become more sophisticated and diversified, organizations are urged to constantly adopt and update measures for contrasting different types of attacks. Particularly, as novel techniques (e.g., social engineering and phishing) are aimed at leveraging individual users' vulnerabilities to attack and breach a larger system or an entire company, user awareness and behavior have become key factors in preventing adverse events, mitigating their damage, and responding appropriately. As a result, the concept of Cyber Hygiene (CH) is becoming increasingly relevant to address the risk associated to an individual's CS practices. Consequently, self-assessment tools are becoming more important for evaluating user's literacy, implementing measures (e.g., training), and studying the effectiveness of interventions. In this paper, we propose a framework for including human factors in the design of self-assessment tools and for accurately modeling CH aspects that the root cause in CS issues.

Keywords: Cybersecurity · Human factors · Phishing · Social engineering · Risk assessment · Cyber Hygiene · Knowledge-Attitude-Behavior

1 Introduction

In the last decade, the widespread adoption of personal communication technology and connected devices changed the scenario of CS: despite the increasing effort of companies and governments to prevent breaches and protect critical business information and organization resources, novel types of CS threats (e.g., ransomware, phishing, and social media engineering) directly aimed at exploiting individuals pose new challenges for entire organizations [1]. In the recent years, most of the work focused on enforcing security of cyber-physical systems aimed at protecting the entire organization: unfortunately, this is not enough to prevent breaches caused by incorrect behavior of their employees and users. Several recent events demonstrated that traditional CS frameworks, including the development of guidelines especially designed to instruct users about their CS practices, are not enough to prevent attacks that directly target individuals via indirect methods (e.g., social engineering) and threats (e.g., phishing and

I. Corradini et al. (Eds.): AHFE 2020, AISC 1219, pp. 88–94, 2020.
https://doi.org/10.1007/978-3-030-52581-1_12

ransomware) that leverage poor adherence to good security practices. Also, as detailed by research studies and incident reports, users are prone to making mistakes and to reiterating incorrect CS behavior, such as reusing the same password for several accounts and generating weak usernames and passphrases, which creates entry points for hackers and, consequently, weaken or void any security measures taken [2, 3]. Therefore, companies are increasingly diversifying CS strategies: awareness campaigns, required training, and informational events, which were demonstrated to lead to a better understanding of the risks and how to avoid them. As users and their practices are the last line of defense and, simultaneously, the first entry-point of the most dangerous attacks [4], the concept of CH, that is, user's CS behavior with specific regards to practices that can increase risk for others, is gaining interest among CS organizations [5, 6]. Particularly, self-assessment questionnaires can be utilized to identify items in which users lack knowledge or are prone to misbehavior and, consequently, design interventions aimed at increasing their awareness.

In this paper, based on previous literature, we introduce a new model that takes into consideration human factors to exactly identify the root cause of individuals' malpractices. By doing this, we aim at supporting the development of initiatives that specifically target the characteristics of the single user, and thus, could lead to better outcomes.

2 Related Work

In addition to protecting their systems, organizations have begun to address security concerns caused by individuals' weaknesses, by implementing training programs aimed at improving the awareness of their employees, with the objective of reinforcing their ability to recognize, avoid, and report threats. Although most studies in the literature have demonstrated the effectiveness of training initiatives in increasing individuals' CS literacy, the authors of [6] found that in several cases previous training does not result in any significant improvement in terms of adoption of more secure practices, unless the strategy, design, and delivery of courses are aligned with assessment policies that enforce correct CS conduct on a continuous basis. As discovered by [7], different factors might influence individual's behavior, which, in turn, makes it difficult to define a holistic framework for addressing the diverse aspects that contribute to neutralizing threats.

The concept of CH aims at introducing a new approach in CS that combines established practices from healthcare domain to refer to individual's CS posture [5, 6]. As experts are still shaping its scope, in this paper, we adopt the definition of [5], that describes CH as *the cyber security practices that online consumers should engage in to protect the safety and integrity of their personal information on their Internet enabled devices from being compromised in a cyber-attack*. Both the definition and its explanation are especially effective in assimilating preventive approaches in CS to measures adopted in healthcare standards. Also, [5] proposed the Cyber-Hygiene Inventory (CHI), that is, a model that enables categorizing questions regarding several items of concern into standard risk dimensions, such as storage and device (S), authentication and credentials (A), Facebook and social media (F), e-mail and messaging (E), and transmission and browsing (T). In general, easy-to-adopt tools in the form of questionnaires and surveys are convenient and versatile instruments for assessing, screening, and monitoring individual's practices on a regular basis, eliciting potential risks, and addressing them with appropriate follow-up interventions. Unfortunately, they lack longevity and require continuous updates to cope with the constantly changing scenario of CS. Conversely, the top-down design of the CHI and its abstraction level render it more robust compared to other questionnaires and scales: the specific risk factors can be further customized to take into account new threats and to change the depth, scope, and content of questions based on the context of application. However, the current design of the CHI only tests subject's knowledge about CS without considering any human factors, such as, behavior, attitude, perception, other intrinsic and extrinsic aspects (e.g., gender, age, and facilitating conditions) that have been demonstrated to be crucial in implementing effective measures. Consequently, the questionnaire offers very limited insight on individual's general attitude with respect to CH as well as on their actions and situational responses in presence of a potential threat. For instance, as showed by previous research, despite knowing how to generate secure passwords and being aware of the risks of reusing the same username for multiple websites, individuals might decide to compromise their standard to a level that results in better convenience [2, 3, 4]. Similarly, the CHI does not support identifying the underlying factors impacting individual's intention to implement a correct behavior, despite of their general attitude to CS practices. As a result, two users assessed with a questionnaire designed using the current CHI model could very well obtain the same CH profile despite their actual actions might result in very different outcomes in terms of risk. For instance, users could adopt strict measures in regard to sharing information via social media, because they take their privacy into consideration for reasons that are not related to any potential implications in terms of CS. Moreover, the current CHI lacks aspects that support updating questionnaires and incorporate new questions aimed at measuring individuals' progress over time and analyze the impact of CS interventions on their CH posture.

3 Incorporating Human Factors in the Inventory

In this paper, we propose a novel framework for designing self-assessment tools in the context of CH. Specifically, our work aims at improving the inventory described in [5], so that human factors can be taken into consideration in the CHI in designing questions, administering assessment tools, and designing interventions. To this end, in addition to the risk contexts considered by [5], our work incorporates the Knowledge-Attitude-Behavior (KAB) model introduced by [8] to address risk in the healthcare domain, which was not included in the original version of the CHI. The KAB approach has been utilized in several CS frameworks and self-assessment scales in the context of CS, such as the Human Aspects of Information Security Questionnaire (HAIS-Q) [9].

3.1 The Importance of Knowledge, Attitude, and Behavior

In our work we expand the KAB model to better fit the concept of CH: as shown in Fig. 1, we define *knowledge* as subject's level of training and awareness of the risk concerning specific aspects in the CS field, that is, technical competence (e.g., authentication and credentials) that is already taken into consideration in most questionnaires, including the CHI; we use *attitude* to refer to individual's general approach to CS based on their perceived level of severity and to recurring patterns in their habits; finally, we consider *behavior* any aspect related to their situational response, that is, the security score associated with actual actions realized by users in order to prevent or address threats in the dimensions considered by the CHI. By doing this, we separate CS concerns related to human factors into three specific domains and, thus, we make it possible to precisely identify the areas in the process that are more prone to potential flaws, so that they can be addressed with targeted intervention. As a result, an individual's CH profile can be obtained by evaluating the risk contexts in combination with human factors. This, in turn, facilitates designing CH inventories that better integrate within a workflow where an improved and more accurate assessment of an individual's CH score results in the prescription of CH interventions especially targeted at the root cause of the CS concern. Furthermore, categorizing risk dimensions into their atomic components is expected to enhance evaluating the effectiveness of CH interventions.

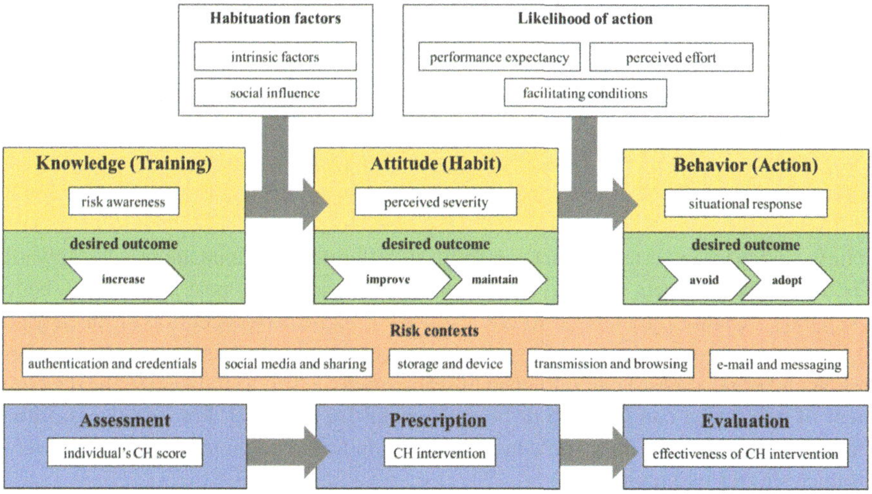

Fig. 1. An overview of our proposed framework. We incorporate the KAB model in the design of CH self-assessment tools to highlight the importance of accounting for human factors that have an impact on individuals' risk awareness, perception of the severity of threats, and on their situational response in different contexts. Also, our framework includes aspects that influence user's attitude and behavior and suggests specific dimensions that need to be considered when designing questionnaires and updating them by considering the desired outcome in terms of CH improvement in the user posture.

3.2 Desired CH Outcomes

Moreover, in our framework we divide each component of the KAB model into specific actions that individuals are expected to realize depending on their current and desired level of CH. Particularly, we take into consideration if they are able to: (1) *increase* their knowledge and awareness, (2) *change* an incorrect attitude in terms of CS so that they can build and (2) *maintain* a high CH profile, and (3) *avoid* potentially dangerous actions and *adopt* strategies that are expected to result in a proactive and improved behavior with respect to detecting and reporting potential threats. As a result, in addition to providing insight on the root cause analysis of CH issues, our framework supports updating data collection instruments by designing questions that specifically evaluate users' growth over time in terms of their level of CH.

3.3 Human Factors Influencing Knowledge, Attitude, and Behavior

Indeed, surveys and questionnaires, such as the CHI, are designed with a two-fold purpose, that is, (1) screening individual's and identifying their CH posture to prevent risk associated with their CS behavior, and (2) measuring the effectiveness of CS interventions, such as training programs, which have the objective of enhancing individuals' knowledge, in order for an increased awareness habituates them to correctly align their perception of risk with its actual severity, so that they can adopt a correct behavior when realizing their tasks. Nevertheless, several intrinsic and extrinsic human factors play a crucial role in the KAB model and they have an impact in correctly translating CS knowledge (risk awareness) into a responsible attitude; also, they intervene in specific situations in which users are required to take appropriate actions. Therefore, our model takes into consideration *habituation factors*, that is, aspects that shape users' habits over time and impact their attitude; also, we suggest elements that modify individuals' *likelihood of action* in accordance with the expected CH behavior. The former includes background, beliefs, and prior experiences that can result in different attitudinal approaches towards CH. For instance, users who have experienced a breach in the past might show better CH habits, because the incident might result in an increased perception of the severity of CS threats. Also, social factors, such as practices adopted by user's groups (e.g., their milieu and organization) can influence the individuals' attitude and lead to the development of habits that can improve or be detrimental for their CH. For instance, limiting the maximum length of passwords in authentication forms might induce users to always produce shorter passphrases.

Furthermore, additional human factors intervene when users actually realize actions (e.g., opening an attachment, changing the privacy settings of their social media account, and creating a password): their behavior can be influenced by the perceived effort and the expected performance, which refer to the difficulty and to the benefits (in terms of CH) of adopting a secure behavior in accomplishing a task, respectively. Moreover, external factors can have an impact on users' actions: facilitating conditions refer to elements that make it easy for users to implement CS principles (e.g., tools for reporting spam and phishing emails), enforce their correct behavior (e.g., scheduled antivirus updates), and prevent potentially dangerous actions (e.g., requiring user's confirmation before opening a suspicious file). For instance, the trade-off between convenience and strength in password creation represents the relationship between performance expectancy and perceived effort; conversely, an example of facilitating conditions is the automatic password generation provided by certain browsers that proactively suggest and memorize secure passphrases without any overhead for the user.

In our model, we consider factors related to social influence as different from aspects pertaining to facilitating conditions: the former refers to practices that result in attitudinal changes (e.g., as users become familiar with password management systems, they develop the habit of using it for every passphrase), whereas the latter indicates adoption of policies or instruments that prevent users from adopting an incorrect behavior by making it more convenient to adopt secure measures (e.g., requiring users to change their passwords every three months or adopting two-factor authentication systems).

4 Conclusions and Future Work

As reports found that a large number of attacks leverage vulnerabilities at the individual user level and use them as entry points, organizations are increasingly adopting assessment tools that help them evaluate individual's awareness with respect to CS and implement interventions, such as training programs, aimed at addressing risk proactively by improving the CH profile of their users.

In this paper, we introduced a novel framework that takes into consideration relevant human factors that have an impact on individual's CH. By doing so, we aim at enhancing the design of self-assessment tools, so that organizations can create better questionnaires that achieve a more in-depth picture of the respondent and enable identifying the types of threats together with their root causes. To this end, we modeled the underlying behavioral aspects that influence user's motivation in perceiving and addressing CS risks properly. The advantage of our model is that, in addition to risk contexts, it suggests dimensions that have to be taken into consideration without detailing individual questions or specific implementation details, which makes it consistent with the strategy adopted by [5] for developing their CHI. This is to maintain a top-down approach that supports evaluating human aspects that are associated with habituation factors and with the likelihood of action separately from users' knowledge about risk items.

References

1. Caporusso, N., Chea, S., Abukhaled, R.: A game-theoretical model of ransomware. In: International Conference on Applied Human Factors and Ergonomics, pp. 69–78. Springer, Cham, July (2018)
2. Stainbrook, M., Caporusso, N.: Convenience or strength? aiding optimal strategies in password generation. In: International Conference on Applied Human Factors and Ergonomics, pp. 23–32. Springer, Cham, July 2018
3. Stainbrook, M., Caporusso, N.: Comparative evaluation of security and convenience trade-offs in password generation aiding systems. In: International Conference on Applied Human Factors and Ergonomics, pp. 87–96. Springer, Cham, July 2019
4. Fandakly, T., Caporusso, N.: Beyond passwords: enforcing username security as the first line of defense. In: International Conference on Applied Human Factors and Ergonomics, pp. 48–58. Springer, Cham, July 2019
5. Vishwanath, A., Neo, L.S., Goh, P., Lee, S., Khader, M., Ong, G., Chin, J.: Cyber hygiene: the concept its measure and its initial tests. Decis. Supp. Syst. **128**, 113160 (2020)
6. Cain, A.A., Edwards, M.E., Still, J.D.: An exploratory study of cyber hygiene behaviors and knowledge. J. Inf. Secur. Appl. **42**, 36–45 (2018)

7. Neigel, A.R., Claypoole, V.L., Waldfogle, G.E., Acharya, S., Hancock, G.M.: Holistic cyber hygiene education: accounting for the human factors. Comput. Secur. **92**, 101731 (2020)
8. Bettinghaus, E.P.: Health promotion and the knowledge-attitude-behavior continuum. Prevent. Med. **15**(5), 475–491 (1986)
9. Parsons, K., Calic, D., Pattinson, M., Butavicius, M., McCormac, A., Zwaans, T.: The human aspects of information security questionnaire (HAIS-Q): two further validation studies. Comput. Secur. **66**, 40–51 (2017)

Habituation: A Possible Mitigation of a Wicked Problem

Kirsten E. Richards(✉)

United States Naval Academy, Annapolis, USA
krichard@usna.edu

Abstract. A construct for intentional habit formation is suggested as a possible mitigation to the disparity between user capability and systems requirements. The importance of usable security is well represented in early discussions ([3]; Sasse 2001). Twenty years after M. S. Ackerman [7] provided a significant discussion of the "gap" between what humans need and what computers can support, the "social-technical gap" in privacy and security management continues. Humans, for many reasons, cannot make good, consistent decisions regarding security. Current and foundational theoretical understandings of human limitations are outlined, in both an individual and social context. The difference between current systems and principles of interface and interaction design are highlighted. Finally, a possible ameliorating step is suggested. Specifically, a movement from reliance on human cognition and decision making to a reliance on habit formation.

Keywords: Usable security · Usable privacy · Human behavior · Human computer interaction · Cyber security

1 Introduction

Human error is consistently cited as the most common cause of security breaches [1, 2]. An increasing awareness of the necessity for usable security has driven many excellent studies and the security community movement away from a "stupid user" mentality [3]. Security requirements and human behaviors are often viewed in tension or as an intractable problem that will continue to exist in a paradoxical state. Even the nature of human needs is often seen as paradoxical – for instance, the framed "privacy paradox" – the tension between user's stated interest in privacy and their behavior [4, 5]. Humans live many parts of our lives in, if not a paradox, at least balancing seemly conflicting needs. To support security, secure behaviors must be consistent. One lapse in judgement or hasty assumption can cause a significant security problem.

Habituation, rather than knowledge training, may be as a possible mitigating approach to assist users in behaving more consistently in line with their own goals or goals of organizations. The formation of security habits will require the implementation of design science principles, particularly consistency [6]. The following sections outline the literature on humans and security, highlighting the difficulty humans face in behaving securely online. Habituation is suggested as possible mitigation to the difficulty users experience.

I. Corradini et al. (Eds.): AHFE 2020, AISC 1219, pp. 95–101, 2020.
https://doi.org/10.1007/978-3-030-52581-1_13

2 Humans and Security

HCI continues to advance rapidly and is showing increased specialization and expansion into developing areas of inquiry as well as relevance to an expanding number of fields as technology becomes more ubiquitous. The difference between the flexibility and nuanced nature of human decision making and brittle systems is well established [7]. People struggle with security, in part, because of the limitations of the design and because of the discrepancy between their strengths and the systems requirements. Privacy decisions are also difficult and form an important part of many cyber-based attacks [8, 9].

2.1 Human Limitations

Vannevar Bush's publication of "As We May Think" heralded a significant development in cognitive models of computer human interaction [10]. Bush described ideas, particularly *information overload* that are highly pertinent to current discussions in security awareness. Furthermore the idea of *associative indexing* describes, with a prophetic degree of accuracy, our current interactions on the web [10, 11]. Through this technology, people are introduced larger cognitive space, where human decision making is hampered by cognitive limitations.

Bounded Rationality. Ackerman [7] highlights the difference between human ability and what systems demand of their users in the need for users manage a near "infinite information space" (p. 186). If anything, the space has grown in the last twenty years and the ability of humans to control their personal data has not. The concept of *bounded rationality* states that humans have certain inherent limitations, both in their ability to calculate the outcomes of their decisions, the limits of the information that they have, and the limited amount of time individuals have to make decisions [12].

Users make decisions with limitations on the amount of time and information [13]. The privacy calculus model highlights the inherent difficultly users face in actually making privacy decisions [14]. Interaction and interface design recommendations highlight the need for simplicity [15] and reduce cognitive load [6]. Infinite space, of unknown but lasting duration does not align with users' decision-making models.

Memory. Human memory is unreliable. Usability research highlights the need for recognition, rather than recall, and cautions designers to avoid overly taxing short-term memory [6, 16]. Miller [17] popularized, "the magical number seven", plus or minus two, as description of limitations humans experience in information processing. Memory is fallible, and short-term memory is at a premium, particularly when a human is engaged in attempting to complete a task, which is addressed more thoroughly later.

Artifacts. Ackerman [7] highlights the strong relationship in design sciences between artifacts, study and theory [18]. In the context of the discussion on bounded rationality, it becomes apparent that there are very few useful physical artifacts. Physical security artifacts are highly bound to their contextualized, physical environments.

There does not seem to be an analogous physical artifact for a near infinite, persistent space. The closest possible analogy seems to be "public" which is not how users perceive all online spaces [19]. Even if users did view all cyber spaces as public spaces, they would be unlikely to be able to control the collection and use of substantial portions of data about themselves or perceive how that data might be used [19]. Lederer, Dey [20] suggest the concept of *faces* as metaphor for privacy management. However, in an interconnected space, various *faces* may be connected, resulting in the opposite of the intention. Recent work are addressing this problem, working to make data in this space more comprehensible to users [21].

Co-evolution. People and their systems are subject to a process of co-evolution, in which sociotechnical systems evolve together towards a state that reduces gaps between requirements and feasibility [22]. The feedback loop is not closely rapidly. Companies and other entities benefit from gathering personal data [23]. Current systems are rigid [7] and that rigidity is intentional. Systems specifications require a certain level of rigidity. The rigidity of systems will require concentrated effort over time to relax. The need for improved security is immediate.

2.2 Social Limitations

Users do not perform security tasks in isolation. Perhaps users are trying to reset a password or respond to an email from a colleague promptly. Successful attacks are often modeled on experiences. The exotic, and unlikely, Nigerian prince is replaced by increasingly convincing and believable emails that mimic our daily experiences [24, 25]. Additional contexts as significant to understanding security choices.

The *Other* Humans. Humans other than end users can contribute to end user errors. For instance, when information systems personnel (authorized users) send legitimate requests to users to update sensitive information via a link. While the email is perfectly legitimate, it also has the potential to habituate users to responding to links asking for sensitive personal information. Instructions to users to ignore security warnings is not uncommon and builds conflicting habits in users. Documented examples of this type of behavior are readily available [26]. Admittedly, the systems administrators and information systems staff frequently do a very good job of finding ways to communicate requirements to users, but consistency in desired user behavior is imperative.

Another example of end user behavior influencing users is the process of conducting business. Halevi, Memon [27] illustrated that increased conscientiousness, with a lower risk perception, equated to an increased success in spear phishing attacks. Very conscientious employees are great, except when their consciousness drives the completion of a task that compromises security.

The *Other* Needs. Security is often a secondary task. For example, a security warning pops up when browsing the web, or a computer needs to be updated, but only at the most inconvenient time. When users are making security and privacy decisions, they may not be thinking about security and privacy, but about their other task, requirement or perceived reward [13, 28]. Maslow and Lewis [29]'s hierarchy of needs includes the need for safety, but also the need for love and belonging, highlighted by "friendship, family, intimacy and sense of connection" (p. 370). It is to this need that a great deal of online information sharing appeals. Users perceive Facebook as a way to receive social support [30].

2.3 Examples of Limitations and Requirements

There are many concrete examples of the cognitive dissonance and influence of the interplay between *self* and *others* inherent in current systems. In the context of security and security decision making, passwords and privacy policies are particularly pertinent examples.

Passwords. The limitations of human memory severely construct the ability of users to perform human requirements for security. There is perhaps no better illustration of this than password requirements. *If a set of rules were constructed to make security as unusable as possible, password requirements could hardly be surpassed.* Random, frequently changed, long, and highly variant, virtually no plan could guarantee a usability disaster. This disparity is highlighted frequently in the literature [31–33].

Privacy Policy. Privacy policies also perfectly illustrate human limitations. Users who claim to be concerned about privacy do not read privacy policy [13, 34]. However, it is worth considering whether or not it would even be possible for users to realistically read privacy policy. McDonald and Cranor [23] explored the amount of time it would take to read privacy policy. They determined individual reading time to be between 181 h per year and 304 h per year, a nearly impossible task [23]. Interface design and interaction design sciences have long relied on certain principles to improve usability. Among these, the need for consistency and simplicity are paramount [6, 16]. Security and privacy settings, policy and user end decision making are neither consistent nor simple.

3 Habits - Mitigation Through Human Strengths

Current technologies rely on areas the exercise of human overcoming significant cognitive, and psychosocial limitations to perform "correct" security actions. Mitigation measures certainly include improving usability but might also include an adjustment in human mechanisms for behavior.

Habituation. One possible solution to the problem of bounded rationality is the implementation of habituation as regards security related behaviors. Many training programs fail to produce lasting results for a variety of reasons, one of which is lack of continuous feedback [35]. Habituation is a viable alternative to repeated decision making. Users could be "trained" not to make decisions, but to develop secure habits. Habits are an "automatic behavioural responses" formed through a cue-reward cycle [36, 37]. Habits are also highly contextual [36]. *Users are currently habituated to bad security habits as a direct result of the impossibility of achieving "good" security practices.* They cannot realistically read privacy policy on websites; so they have the very bad habit of ignoring privacy policies. They cannot, realistically, perform "well" with passwords, therefore, they select a "good password" rule to break.

Habits are "cued by context" [36], meaning that a variety of security habits could be formed based on context. Since habits are contextual, habits also represent a flexible state that also bypasses complex decision making. *Habits could be a bridge between the inherent flexibility of human decision making and the inherent inflexibility of systems.*

Habits also require less cognitive engagement. Habits may also be intentionally formed in relationship to goals [36] and changing habits is difficult, but not impossible. Habits appear to be a human version of the simplification and rigidity we observe in systems. Habits are also responsive to external stimuli [36]. Therefore, those interested in forming specific security habits, such as employers, may use habituation, rather than cognitive training, to form very strong contextual responses that will reflect the security needs of the organization [36].

4 Conclusion

Limitations, such as *bounded rationality* and limited *information processing capacity* make managing security in the current context virtually impossible for users. However, there are ameliorating factors, already well applied to other problems in this space that can be implemented in security settings. The suggestion presented here is replacing constant decision making with *habituation* facilitated by applying principles of *consistency.* Humans cannot operate effectively in the systems space we have created. Purposeful design around the concept of security habits require the implementation of key design principles and strict adherence to these principles on the part of designers. Habituation could allow humans to behave in complex ways in the context of the difficult decision making spaced posed by system.

References

1. Swanson, M., Guttman, B.: Generally accepted principles and practices for securing information technology systems. National Institute of Standards and Technology, Technology Administration (1996)
2. Ahmed, M., et al.: Human errors in information security. Int. J. Adv. Trends Comput. Sci. Eng. **1**(3), 82–87 (2012)
3. Adams, A., Sasse, M.: Users are not the enemy. Commun. ACM **49**(12), 41–46 (1999)
4. Kokolakis, S.: Privacy attitudes and privacy behaviour: a review of current research on the privacy paradox phenomenon. Comput. Secur. **64**, 122–134 (2017)
5. Norberg, P.A., Horne, D.R., Horne, D.A.: The privacy paradox: personal information disclosure intentions versus behaviors. J. Consum. Affairs **41**(1), 100–126 (2007)
6. Shneiderman, B., et al.: Designing the User Interface: Strategies for Effective Human-Computer Interaction. Pearson Education, London (2016)
7. Ackerman, M.S.: The interllectual challenge of CSCW: the gap between social requirements and technical feasibility. Hum.-Comput. Interact. **15**(2–3), 179–204 (2000)
8. Richards, K.E.: Risk analysis of the discoverability of personal data used for primary and secondary authentication. University of Maryland Baltimore County, MD, USA (2017)

9. Reeder, R., Schechter, S.: When the password doesn't work: secondary authentication for websites. IEEE Secur. Priv. **9**(2), 43–49 (2011)
10. Bush, V.: As we may think. Atlantic Monthly, pp. 101–108 (1945)
11. MacKenzie, I.S.: Human-computer interaction: An empirical research perspective. Elsevier, New York (2013)
12. Simon, H.A., Bounded rationality. In: Utility and probability. pp. 15–18. Springer (1990)
13. Acquisti, A., Grossklags, J.: Privacy and rationality in individual decision making. IEEE Secur. Priv. **3**(1), 26–33 (2005)
14. Dinev, T., Hart, P.: An extended privacy calculus model for e-commerce transactions. Inf. Syst. Res. **17**(1), 61–80 (2006)
15. Thomas, J.C., Richards, J.T.: Achieving psychological simplicity: Measures and methods to reduce cognitive complexity. Hum.-Comput. Interact.: Des. Issues Solut. Appl. **161**, 489–508 (2009)
16. Nielsen, J.: Ten usability heuristics (2005). http://www.nngroup.com/articles/ten-usability-heuristics/. Accessed
17. Miller, G.: The magical number seven, plus or minus two some limits on our capacity for processing information. Psychol. Rev. **101**(2), 343–352 (1955)
18. Olson, G.M., Olson, J.S.: Research on computer supported cooperative work. In: Handbook of Human-Computer Interaction, pp. 1433–1456. Elsevier (1997)
19. Tan, Q., Pivot, F.: Big data privacy: changing perception of privacy. In: 2015 IEEE International Conference on Smart City/SocialCom/SustainCom (SmartCity). IEEE (2015)
20. Lederer, S., Dey, A.K., Mankoff, J.: Everyday privacy in ubiquitous computing environments. In: Ubicomp 2002 Workshop on Socially-Informed Design of Privacy-Enhancing Solutions in Ubiquitous Computing (2002)
21. Wilkinson, D., et al.: Privacy at a glance: the user-centric design of glanceable data exposure visualizations. Proc. Priv. Enhancing Technol. **2020**(2), 416–435 (2020)
22. Moor, A., Aakhus, M.: Argumentation support: from technologies to tools. Commun. ACM **49**(3), 93–98 (2006)
23. McDonald, A.M., Cranor, L.F.: The cost of reading privacy policies. J. Law Policy Inf. Soc. **4**, 543 (2008)
24. Wen, Z.A., et al.: What. hack: learn phishing email defence the fun way. In: Proceedings of the 2017 CHI Conference Extended Abstracts on Human Factors in Computing Systems (2017)
25. Richtel, M., Kopytoff, V.G.: E-mail fraud hides behind friendly face. The New York Times, p. 2 (2011)
26. Zurko, M.E.: User-centered security: stepping up to the grand challenge. In: 21st Annual Computer Security Applications Conference (ACSAC 2005). IEEE (2005)
27. Halevi, T., Memon, N., Nov, O.: Spear-phishing in the wild: a real-world study of personality, phishing self-efficacy and vulnerability to spear-phishing attacks. Phishing Self-Efficacy and Vulnerability to Spear-Phishing Attacks, 2 January 2015
28. Alashoor, T., Al-Maidani, N., Al-Jabri, I.: The privacy calculus under positive and negative mood states (2018)
29. Maslow, A., Lewis, K.J.: Maslow's hierarchy of needs. Salenger Inc. **14**, 987 (1987)
30. Olson, D.A., Liu, J., Shultz, K.S.: The influence of Facebook usage on perceptions of social support, personal efficacy, and life satisfaction. J. Organ. Pscol. **12**(3/4), 133–144 (2012)
31. Bonneau, J., et al.: The quest to replace passwords: A framework for comparative evaluation of web authentication schemes, pp. 553–567 (2012)
32. Brown, A.S., et al.: Generating and remembering passwords. Appl. Cogn. Psychol. **18**(6), 641–651 (2004)
33. Grawemeyer, B., Johnson, H.: Using and managing multiple passwords: a week to a view. Interact. Comput. **23**(3), 256–267 (2011)

34. Vila, T., Greenstadt, R., Molnar, D.: Why we can't be bothered to read privacy policies models of privacy economics as a lemons market. In: Proceedings of the 5th International Conference on Electronic Commerce (2003)
35. Bada, M., Sasse, A.M., Nurse, J.R.: Cyber security awareness campaigns: Why do they fail to change behaviour? arXiv preprint arXiv:1901.02672 (2019)
36. Wood, W., Neal, D.T.: A new look at habits and the habit-goal interface. Psychol. Rev. **114**(4), 843 (2007)
37. Lally, P., Gardner, B.: Promoting habit formation. Health Psychol. Rev. **7**(sup1), S137–S158 (2013)

Developing Digital Awareness at School: A Fundamental Step for Cybersecurity Education

Isabella Corradini[1,3](✉) and Enrico Nardelli[2,3]

[1] Themis Research Centre, Rome, Italy
isabellacorradini@themiscrime.com
[2] Department Mathematics, University Roma Tor Vergata, Rome, Italy
nardelli@mat.uniroma2.it
[3] Link & Think Research Lab, Rome, Italy

Abstract. The theme of cybersecurity regards people *in primis*, considering that everyone uses digital technologies both in professional and private life, and that people's behaviour plays an important role in the occurrence of cyberthreats. The human factor has therefore to be recognized as an essential element to be considered for developing an effective cybersecurity, and education is the key driver. However, since children access online activities at an early age, it is wise to develop interventions to promote digital awareness from first years at school, focusing on the responsible use of digital technologies. Becoming conscious of the risks they are exposed to is an important step for children to move safely on the Internet and to understand the different cyber-risks they have to face. This activity represents hence a fundamental step for cybersecurity education.

In this paper we present a study investigating Italian school teachers' perception of their students' digital awareness and their evaluation of the actions needed for its development. Answers were provided by 2,229 teachers from all over the country belonging to primary and secondary schools, participating in a national project whose goal is to spread computer science and to sensitize students to a proper use of digital technologies.

The results confirm the high sensitivity of teachers towards digital awareness issues. Indeed, students should be prepared to recognize risks when they use digital technologies: not only cyberbullying, they should pay more attention to the protection of their personal data, and to the reliability of news on social media. Moreover, teachers declare the need for themselves to receive specific training on digital awareness, and to be supported in their activities.

Keywords: Digital awareness · Cybersecurity · Education

1 Introduction

Over the last few years, the theme of cybersecurity has become a great challenge for every country and a significant problem to handle for all organizations. Cyber-threats

I. Corradini et al. (Eds.): AHFE 2020, AISC 1219, pp. 102–110, 2020.
https://doi.org/10.1007/978-3-030-52581-1_14

and attacks continue to grow, notwithstanding the availability of innovative technological solutions and important regulations aimed at protecting personal information (The General Data Protection Regulation, GDPR).

Many are convinced that cybersecurity is above all a human problem (e.g. [1, 2]), and that cybersecurity awareness programmes are needed to respond to the gap between people and digital technology [3]. However, there is still a general tendency to mainly focus on a technical perspective, thinking that innovative solutions, for example the newest ones based on Artificial Intelligence, are able to solve both current and future security problems.

On the other hand, developing a different attitude towards cybersecurity is not easy, since it requires an ongoing process based on awareness and education programmes. Outcomes are not immediate, but it is now evident that the poor results deriving from current cybersecurity approaches impose a different intervention model, where human beings must be considered as an essential part of the solution [4]. In fact, if the problem is mainly represented by unsecure behaviour, it is fundamental that people are fully aware of cyber-risks and respond to them appropriately. Training and education are the key drivers [5].

Considering that everyone uses digital technologies for both professional and private life, and that children access online activities at an early age and spend more and more time in using digital technologies [6, 7], it is wise to develop interventions to promote digital awareness at school [8]. Children often are not aware of the risks they are exposed to [9], for instance sharing personal photos and posting sensitive information on social media. Moreover, we should not forget that also schools can be a target of cyberattacks [10].

If on one side there is a wide interest in developing cyber-skills at school for future careers in cybersecurity, on the other side it is important to boost digital awareness early in school. This activity, if well-managed, can represent a primary step in order to become aware of cybersecurity risks. It is clear that, considering the age of students, it is not appropriate to talk about cybersecurity; instead, digital awareness or "digital hygiene" might be more respondent to the specific needs, focusing on simple notions about online behaviour. In this sense, it is important to consider that developing digital competences also include soft skills, such as critical thinking, interacting through digital technologies, protecting personal data and privacy [11]. Therefore, increasing awareness of the risks and supporting capacity building of educators in online safety is part of Digital Education Action Plan [12]. These recommendations are fundamental to create educational curriculum to teach computer science at school, considering its social aspects, too. For example, the National Curriculum in England [13] regarding computing programmes of study includes, beyond the scientific and technical aspects of computer science, understanding and using technology, safely, respectfully and responsibly, e.g. recognizing inappropriate contents, protecting online identity and privacy.

In the following, we present a study investigating Italian school teachers' perception of their students' awareness about the use of digital technologies, and the need to be prepared to handle this issue with their students. The study is part of a monitoring report that every year is conducted in order to evaluate general participation of teachers and students in a national project on computer science education [14].

2 Methodology

Programma il Futuro project ([14, 15]) has the goal to increase awareness in Italian schools both on the scientific principles of digital technologies and on the basic concepts for their responsible use. For these goals the project provides lessons developed on the basis of Code.org materials (for awareness on the scientific principle), and guidebooks based on Common Sense materials (for awareness on responsible use of digital technologies). Teachers and students are voluntarily enrolled in the project that, in its fifth year, has involved more than 30,000 teachers and over 2 million students.

Every year teachers are asked to fill out monitoring questionnaires consisting of 40 items to evaluate their general participation – together with students - in project activities, and the quality of the actions implemented.

One section of the monitoring questionnaire, "Digital Awareness", aims at investigating teachers' perception regarding the responsible use of digital technologies by their students and the level of usefulness of the related teaching material developed by the project. The section consists of 15 multiple-choice questions and one open-ended question.

Three areas have been investigated:

Area 1: ***Assessment of the usefulness of digital awareness guidebooks***. Teachers are asked to evaluate, using a scale from 1 (low) to 4 (high) how useful each of the following guidebooks provided by the project is to develop digital awareness in students:

- Super Digital Citizen;
- The Power of Words;
- Private and Personal Information;
- Safe Online Talk;
- Going Places Safely;
- Follow the Digital Trail;
- Screen Out the Mean.

These guidebooks, intended for teacher use, contain fully developed lesson plans with teaching content and exercises on different issues, for example: how to use the Internet and social network safely, how to safeguard personal data and digital reputation.

Area 2: ***Responsible use of digital technologies***. This area investigates:

- what is necessary to develop a conscious use of digital technologies;
- which activities students mainly carry out through digital technologies (e.g., studying, doing research, getting information, playing music);
- how important the knowledge of certain issues is for students' preparation (e.g., fake news, online behaviour).

Area 3: ***Supporting teachers with specific training in digital awareness and security***. This area investigates what type of activity is useful to support teachers in developing a proper digital awareness among their students (e.g. training, communication). Moreover, teachers are asked to evaluate their need for specific preparation on topics related to digital awareness and security.

The monitoring questionnaire was sent in December 2019, through the project platform. The total sample who filled out the questionnaire is composed by 2,229 teachers, presenting the following demographic characteristics: Gender (F: 82.01%; M: 17.99%); Age (up-to-30: 0.27%; 31–40: 5.83%; 41–50: 34.19%; 51–60: 51.91%; 61-and-up: 7.81%). The gender distribution is in line with the national distribution of teachers' gender. The majority of teachers is from primary school (59.98%), then around a quarter (26.29%) from middle school, and a minority is from high school (10.32%). There is also a very limited participation from kindergarten teachers (3.01%). All respondents have participated in "Programma il Futuro" activities for at least one year (the project has been active since 2014).

3 Results and Discussion

We now report and discuss the main outcomes of our study related to the above three areas of the questionnaire.

Area 1. ***Assessment of the usefulness of digital awareness guidebooks***
Results confirm a high perceived usefulness of the guidebooks used to spread digital awareness (Fig. 1). Note that between 80% and 90% of teachers evaluated the guidebook with "high" or "medium" usefulness. Teachers who still do not know these materials

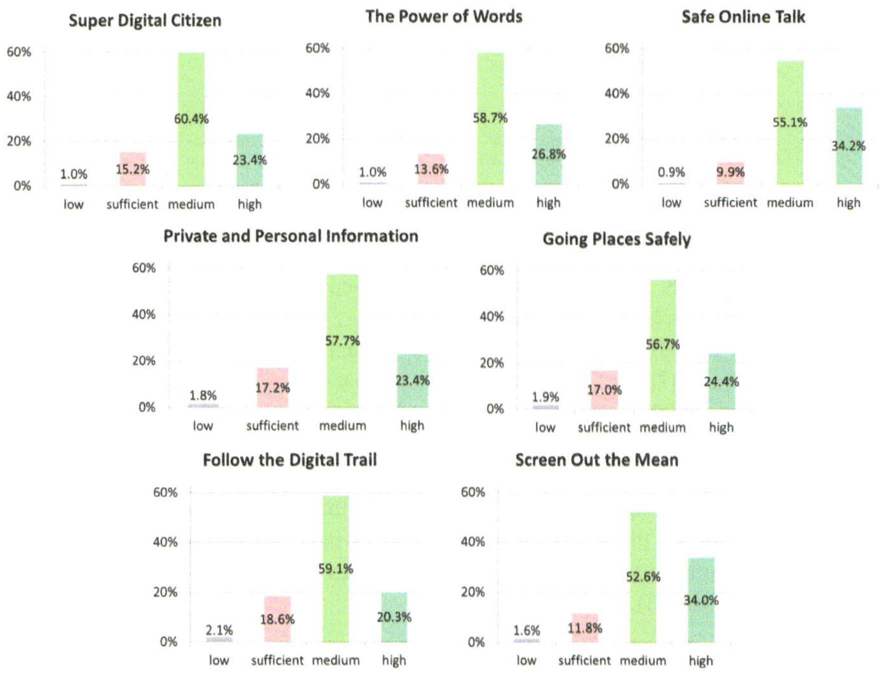

Fig. 1. Usefulness of the various guidebooks for digital awareness.

declare their intention to apply them in their classrooms, because of the topical nature of the issues dealt with. In particular, materials focused on communication (e.g. "Safe Online Talk") have been particularly appreciated by teachers.

The interest demonstrated by teachers is also confirmed by suggestions provided in the open-ended question, where they underline the need for producing further materials on online communication issues, especially about the use of social network.

Area 2 – ***Responsible use of digital technologies***

For a responsible use of digital technologies, teachers think that this mainly requires the development of an adequate knowledge of risks associated with their use (Fig. 2). Therefore, students should be prepared to recognize risks when they use digital technologies. The prevalence of risk element in digital awareness confirms results deriving from previous monitoring report [9]. After "knowledge of risks" (66%), a responsible use of digital technologies passes through "the ability to use them effectively" (49%), "understanding how they work" (45%), and "sense of responsibility" (41%). It was possible to provide up to 3 answers out of 7 and the remaining 3 were selected by less than 40% of respondents.

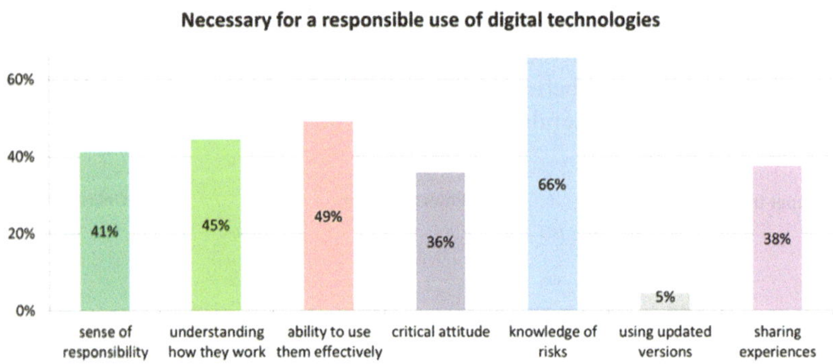

Fig. 2. What is necessary for a responsible use of digital technologies.

For what regards how digital devices are used by students (Fig. 3), according to teachers' perception they use them mostly "to play" (85%), "to communicate/share with friends/classmates" (56%), "to listen/watch/download music" (54%), "to study/research" (28%), and in only 11% of cases "to get information" (multiple answers were possible).

The high value obtained by "playing" agrees with the fact that many participants involved in the study are primary school teachers. Clearly, at this stage, students' activities are not focused on searching information on the Internet. However, it is recognized the social significance that digital tools and devices have for students, given that they permit to communicate with their friends/schoolmates and to share news (in particular in the secondary school).

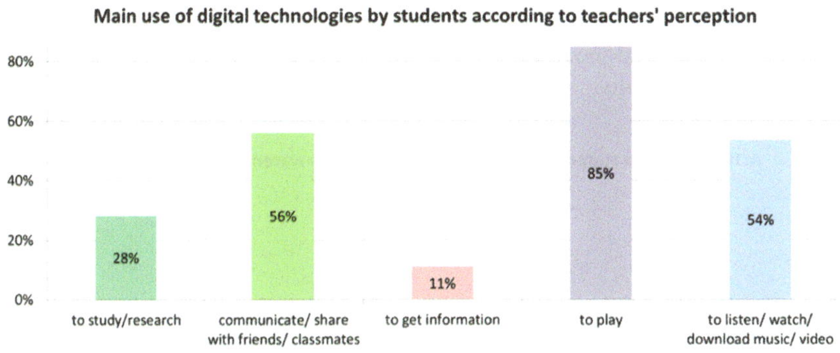

Fig. 3. How students use digital technologies according to teachers' perception.

In Fig. 4 you can see how teachers evaluated the importance of various issues to prepare students to use digital technologies in a safe and responsible way. "On line harassment", "Data protection and privacy", and "Safe online behaviour" are the most important topics for students' preparation. If on the one side "Online harassment" received the highest evaluation, given it is probably the hottest topic in the discussion on the Internet risks regarding children and teenagers, on the other one it is interesting to notice that teachers appropriately recognize the importance of protecting data and guaranteeing privacy while online.

Fig. 4. Degree of importance of various topics for students' digital awareness.

Area 3 – ***Supporting teachers and specific training in digital awareness and security***
Regarding the activities considered useful to help teachers to develop students' digital awareness (Fig. 5), the most important action according to teachers is "sharing experiences" (65%), followed by "training in presence" (53%), and "parents' involvement" (48%). It was possible to provide up to 3 answers out of 7, and the remaining ones were

selected by less than 40% of respondents. These results confirm the importance of dealing with these issues and the need for the contribution of others, e.g. parents' students, to achieve effective results.

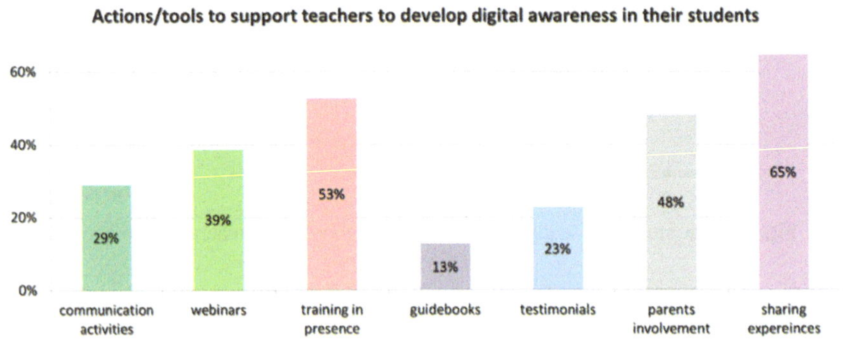

Fig. 5. Actions/tools to support teachers to develop digital awareness in their students.

On the other hand, teachers highlight the need for themselves to receive a specific training on digital awareness issue, as a response to a specific yes/no question (Y: 97.74%). Among the topics to be included in this training they identify (Fig. 6): "social media use" (70%), "fake news" (57%), "online harassment" (56%), and "identity theft" (48%). It was possible to provide up to 4 answers out of 7, and the remaining 3 were selected by less than 45% of respondents.

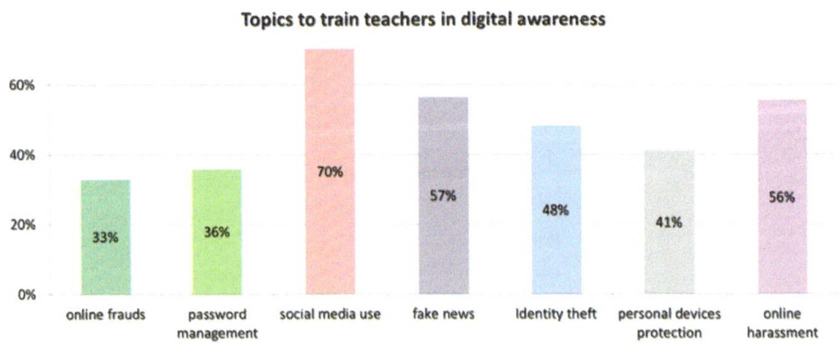

Fig. 6. Topics for training teachers in digital awareness.

This training could provide useful clarifications about common terms used in the area of cybersecurity, given that some of them are often improperly used, because of the lack of knowledge.

4 Conclusions and Future Work

In this paper we have analysed how teachers evaluate the importance of spreading digital awareness among their students and the necessary actions in order to improve the responsible use of digital technologies. Answers from 2,229 teachers - belonging to primary and secondary school - confirm their high sensitivity towards digital awareness issues. Students should be prepared to recognize risks when they use digital technologies: not only cyberbullying, they should pay more attention to the protection of their personal data, as well as to the reliability of news on social media. Becoming conscious of the risks they are exposed to on the Internet is an important step for children to move safely and to understand the different cyber-risks they have to face.

Considering that digitization is unstoppable and that cyberthreats are likely to grow, a precocious education on the responsible use of digital technologies can be a fundamental step for an effective cybersecurity education. Indeed, we think that, given the age of students, talking of cybersecurity training in primary and secondary school is excessive; instead, educating students to understand the concept of digital risks and highlighting the importance of online behaviour can be an effective way to develop digital awareness.

Finally, also teachers need to be supported in this activity; among the preferred actions, they identify sharing of experiences, training classrooms and parents' involvement. They declare the need for themselves of a specific training on digital awareness, and the use of social media is one of the most important topics to be managed.

In conclusion, we think that education is an essential key to handle cyber risks, now and in the future. Not only for workers [5], but also for students and, in general, for all citizens. We will continue to produce guidebooks and materials on the basis of teachers' requests, and to monitor the project activities.

Acknowledgements. We thank all teachers involved in "Programma il Futuro" for their continued participation to the project and their involvement into this research. Thanks also to Common Sense for having provided us the digital awareness teaching material and to Code.org for their activity to support Informatics education in schools. Last, but not least, a big thank to Francesco Lacchia for his excellent work in the adaptation of the Common Sense guidebooks.

References

1. Schneier, B.: Secrets and Lies. Wiley, New York (2000)
2. Safa, N.S., Solms, R.V., Futcher, L.: Human aspects of information security in organisations. Comput. Fraud Secur. **2016**(2), 15–18 (2016)
3. Corradini, I.: Building a Cybersecurity Culture in Organizations. How to Bridge the Gap between People and Digital Technology. Springer, Cham (2020)
4. Zimmermann, V., Renaud, K.: Moving from a "Human-as-Problem" to a "Human-as-Solution" cybersecurity mindset. Int. J. Hum Comput Stud. **131**, 169–187 (2019)
5. Corradini, I., Nardelli, E.: Building organizational risk culture in cyber security: the role of human factors. In: AHFE 2018, pp. 193–202. Springer (2018)
6. Livingstone, S., Haddon, L., Görzig, A., Olafsson, K.: Risks and Safety on the Internet: The Perspective of European Children: Full Findings and Policy Implications from the EU Kids Online Survey of 9–16 Year Olds and Their Parents in 25 Countries. LSE, London (2011)

7. Kardefelt-Winther, D.: How does the time children spend using digital technology impact their mental well-being, social relationships and physical activity? An evidence- focused literature review, Innocenti Discussion Paper 2017-02, UNICEF Office of Research – Innocenti, Florence (2017)
8. Schilder, M.J.D., Brusselaers, B.J., Bogaerts, S.: The effectiveness of an intervention to promote awareness and reduce online risk behavior in early adolescence. J. Youth Adolescence **45**, 286–300 (2016)
9. Corradini, I., Nardelli, E.: Awareness in the online use of digital technologies, In: 11th International Conference of Education, Research and Innovation (ICERI-2018), pp 7036–7042, Sevilla, Spain, November 2018
10. Modan, N.: Recent school ransomware attacks highlight need for ongoing vigilance. EducationDive, January 2020
11. European Commission, The Digital Competence Framework 2.0
12. European Commission, Digital Education Action Plan
13. Gov.UK, Dep. Education, National Curriculum in England: computing programmes of study, September 2013
14. Corradini, I., Lodi, M., Nardelli, E.: Computational thinking in italian schools: quantitative data and teachers' sentiment analysis after two years of programma il futuro project. In: ITiCSE 2017, ACM (2017)
15. Nardelli, E., Corradini, I.: Informatics Education in School: A Multi-Year Large-Scale Study on Female Participation and Teachers' Beliefs. In: ISSEP-2019, pp. 53–67, Cyprus, November 2019

Social, Economical and Behavioral Aspects of Cybersecurity

Economic Prospect Theory Applied to Cybersecurity

Wayne Patterson[1](✉) and Marton Gergely[2]

[1] Patterson and Associates, 201 Massachusetts Ave NE, Suite 316, Washington, DC 20002, USA
waynep97@gmail.com

[2] Department of Information Systems and Security, College of IT, United Arab Emirates University, Al Ain, UAE
mgergely@uaeu.ac.ae

Abstract. A growing concern in the cybersecurity community evaluation of the strengths and defenses regarding cyberattacks. One approach that has not been often explored is to estimate the strength of an attack or defense in economic terms. For example, estimation of the memory required for code used for an attack, or what is equivalent, the computer time to execute an attack. We choose to express the costs in economic terms, and thus define the method of analyzing an important line of research known as "behavioral economics", pioneered by Kahneman and Tversky, and translated into cybersecurity terms. In this way we attempt to determine a cybersecurity analog for well-known results in economic prospect theory to be able to estimate the costs of cyberattacks and defenses.

Keywords: Prospect theory · Cybersecurity · Cyberattacker · Cyberdefender · Kahneman-Tversky

1 Introduction

Over the past 30 years, there has been an emergence of a new approach to economic prospect and utility theory that has contradicted the prevailing views in that discipline. Utility theory stretches back to the work of Jeremy Bentham and others in developing what is now known as utility theory. Only in recent years has the basis of utility theory come under challenge, with some very revealing results about economic choices that have demonstrated using classical assumptions that many results can result in contradictory decisions about such choices.

In this revised approach, many of the hypotheses of what leads to decisions in terms of economic choices have been refuted in this new branch that is usually referred to as "behavioral economics". Perhaps the primary proponents of this new approach have been Daniel Kahneman and Amos Tversky. The latter received in 1985 the Nobel Prize in Economics for his research in this area, and their seminal paper in this regard is "Prospect Theory: An Analysis of Decision Under Risk" published in Econometrica in 1979. We will refer to the results of this paper as "KT".

I. Corradini et al. (Eds.): AHFE 2020, AISC 1219, pp. 113–121, 2020.
https://doi.org/10.1007/978-3-030-52581-1_15

The test data developed in KT were presented in a very general context. As an example, one of the KT questions was "Choose between A: 2500 with probability .33; 2400 with probability .66; 0 with probability .01; 4 B: 2400 with certainty." Raising a question in this fashion gave no information in terms of the meaning of magnitudes of 2500, 2400, or 0 in terms of the values that might be assigned to those magnitudes.

We were interested in analyzing rational choices of individuals in a cybersecurity context. Both from the perspective of a cybersecurity defender and a cybersecurity attacker, decisions to create an attack for a defense can be translated using some metric to an economic value (Patterson and Winston-Proctor 2019).

Consequently, we thought that it would be valuable in trying to understand the motivation of a cyber attacker or defender to be able to assess their decision-making when cast in economic terms; thus it seemed a logical step try to apply the groundbreaking KT research in order to see if their findings that demonstrated weaknesses in classical utility theory might also apply in a cybersecurity environment.

2 Test Design

In this regard, we translated many of the questions posed in the research in Kahneman and Tversky (op. cit.) to a cybersecurity context. The original research carried out by Kahneman and Tversky used subjects at the University of Stockholm, Sweden, and the University of Michigan, Ann Arbor, USA.

In order to carry out this research, we recruited subjects from junior- and senior-level cybersecurity courses at a large university in the United Arab Emirates. All subjects remained completely anonymous and were informed that their participation in the study was entirely voluntary (Tables 1 and 2).

Table 1. Attacker Test Participants (n = 73).

Gender	Male = 19	Female = 54	Age 18–29 = 69
Ethnicity	South Asian = 1	Middle Eastern = 72	Major: Information Systems = 73
Work	Full-time = 4	Part-Time = 2	Not Applicable = 67
Student Status	Freshman 1	Sophomore 12	Junior 23
	Senior 30	Graduate 5	Other 2

3 Preparation and Modification of Test Questions

In the original KT paper, there was no context given for the questions which were used to develop responses from various audiences. In our case, in order to set the context to provide alternative choices, we felt it was necessary to provide some introduction in terms of the description of the cybersecurity environment. Using random assignment, we divided the subjects into two groups, and instructed each group differently, in one

Table 2. Defender Test Participants (n = 69).

Gender	Male = 19	Female = 50	Age 18–29 = 69
Ethnicity	Middle Eastern = 69	Major	Information Systems = 69
Work	Full-time = 5	Part-Time = 4	Not Applicable = 60
Student Status	Freshman 0	Sophomore 13	Junior 16
	Senior 33	Graduate 4	Other 3

case assuming that they would take the point of view of a cyber attacker, and for the other group the role of a cyber defender.

In order for the questions to carry the same magnitude as those found in the original KT questions, we decided to transform the monetary values from those indicated in the original paper, to the currency of the responders, namely UAE dirhams. In that paper, published in 1979, the currency indicated was Israeli pounds, and an average monthly income was approximately 3,000 lb. At the time of this writing, the average monthly family income in the United Arab Emirates was approximately 15,000 dirhams (see https://authorityjob.com/good-average-salary-income-usd-dubai-family/). Thus to have the testees recognize monetary figures of near equivalent values, we decided to multiply all the original values in Kahneman and Tversky (op. cit.) by a factor of 4. Thus in the very first example, Kahneman and Tversky asked for a choice between "choose between (A): 2500 with probability .33, 2400 with probability .66, 0 with probability .01; or (B) 2400 with certainty". In our case, we rewrote this as "(A): 33% chance to win 10000 dirhams, 66% chance to win 9600 dirhams, 1% chance to win nothing; or (B) 100% chance to win 9600 dirhams". In addition, the questions given to the "cyberdefender" students were labelled Dx (x = 1, …, 16), and Ay (y = 1, …, 16) for the "cyberattacker" students.

Similar transformations were made for all questions.

4 Questions for the Current Study

Test questions as reported on in KT were replicated, and in some cases recast in their presentation to address prospect theory questions in a cybersecurity context. In the cybersecurity context, two test subject groups were selected. In one subgroup, the assumption was that the subjects were asked to answer questions assuming they were cyberattackers, and in the other subgroup, that they were cyberdefenders.

In order for the subjects to assess the test questions, they were asked to consider a context for the questions from the perspective of a cyberattacker (for one subgroup), or from the perspective of a cyberdefender (for the other subgroup), as follows:

"As a student of cybersecurity, it is often helpful to imagine you are an attacker or a defender. These questions will give you the opportunity to choose what strategy you would take given each of a number of cyber-attack or defense scenarios.

> "In each, you may assume that your choice will result in an outcome which will yield some value to you (either as an attacker or defender), and that value will

be in accordance with a given probability or set of probabilities. The values will be expressed in dirhams and the probabilities in percentages, where a certainty is 100%."

Further instructions were provided separately for each group essentially defining the activity of a cyber attacker or defender.

The examples for both subgroups, as well as the original KT questions, follow. The numbering of the questions varies slightly but total 16 questions. For the cyber attacker and defender subgroups, the questions are numbered 1-16. For the KT questions, their enumeration is followed: 1–12, 13, 13', 14, 14'.

(A1 and D1) Choose between A and B where you would receive A: 33% chance to win 10000 dirhams; 66% chance to win 9600; 1% chance to win 0 B: 100% chance to win 9600 dirhams
(A2 and D2) Choose between A and B where you would receive A: 33% chance to win 10000 dirhams; 67% chance to win nothing B: 34% chance to win 9600 dirhams; 66% chance to win nothing
(A3 and D3) Choose between A and B where you would receive A: 20% chance to win 16000 dirhams; 80% chance to win nothing B: 100% chance to win 12000 dirhams
(A4 and D4) Choose between C and D where you would receive C: 20% chance to win 16000 dirhams; 80% chance to win nothing D: 25% chance to win 12000 dirhams; 75% chance to win nothing
(A5 and D5) Choose between Choice A and B where you would receive a non-monetary payoff in a given (even though the subjects could perform a conversion to a monetary value), per the corresponding probability A: 50% chance to win a 3-week tour of England, France, Italy; 50% chance: no tour B: 100% chance to win a 1-week tour of England
(A6 and D6) Choose between Choice A and B where you would receive a non-monetary payoff (even though the subjects could perform a conversion to a monetary value), subject to the corresponding probability A: 5% chance to win a 3-week tour of England, France, Italy; 95% chance of no tour B: 10% chance to win a 1-week tour of England; 90% chance of no tour
(A7 and D7) Choose between Choice A and B where you would receive a payoff in a given amount, subject to the corresponding probability A: 45% chance to win 24000 dirhams; 55% chance to win nothing B: 90% chance to win 12000 dirhams; 10% chance to win nothing
(A8 and D8) Choose between Choice A and B where you would receive a payoff in a given amount, subject to the corresponding probability A: 0.1% chance to win 24000 dirhams; 99.9% chance to win nothing B: 0.2% chance to win 12000 dirhams; 99.8% chance to win nothing

(*continued*)

(continued)

(A9) You would like to increase your probability of being able to hack into another computer. You're aware of certain hacking software, but you see an ad for a "probabilistic hacking package" which only costs half as much, but advertises that 50% of the time it will fail in an attack, and the other 50% it will successfully hack. Given this, would you purchase the probabilistic hacking package:
(D9) You are aware of various defender software packages. You see a new ad for "probabilistic virus protection". For this package you pay only half of the regular premium, but the virus protection package only guarantees success 50% of the time. Given this, would you purchase the probabilistic virus protection package:
A: Yes B: No

(A10) Earlier, you attacked an opponent's system, and in this attack, you had a 75% chance of getting caught, and a 25% chance of not being detected, so you could try again. Later, you have a choice between:
A: 50% chance to gain 4000 dirhams; 50% chance of getting caught
B: 100% chance to gain 2000 dirhams
(D10) Earlier, you defended against an opponent, and in this defense, you had a 75% chance of successfully defending, and a 25% chance of being penetrated. On a subsequent attack, you have a choice between:
A: 50% chance to lose 4000 dirhams; 50% chance losing nothing
B: 100% chance to lose 2000 dirhams

(A11) You have already downloaded an opponent's files that you value at 4,000 dirhams. You can continue your hacking search for more assets but only for a limited amount of time. You have two choices:
A: 50% chance to gain 4000 dirhams; 50% chance to lose nothing
B: 100% chance to gain 2000 dirhams
(D11) You have defeated an opponent's attacks that you estimate has cost the opponent 4,000 dirhams. You can continue your defensive tactics that will cost the opponent, but restrict your ability to run other of your software. You have two choices:
A: 50% chance to cost the attacker 4000 dirhams; 50% chance to cost nothing
B: 100% chance to cost the attacker 2000 dirhams

(A12) You have already downloaded an opponent's files that you value at 8,000 dirhams. You can continue your hacking search for more assets but only for a limited amount of time, and you may be detected and lose a certain amount on a reverse hack. You have two choices:
(D12) You have defeated an opponent's attacks that you estimate has cost the opponent 4,000 dirhams. You can continue your defensive tactics that will cost the opponent, but restrict your ability to run other of your software, which will cost you in your computing time. You have two choices:
C: 50% chance to lose 4000 dirhams; 50% chance to lose nothing
D: 100% chance to lose 2000 dirhams

(continued)

(continued)

(A13) You hack into an opponent's account, and you know you only have a limited time before being detected. You may gain some resources according to the following two choices: (D13) As in some earlier questions, if I can detect an intruder and report the intrusion to the local legal authorities, I can likely get a reward for identifying the perpetrator. Since in the limited time for detection, I will only be able to stop one of two attackers. Depending on the value of prosecuting an attacker, my reward is likely to be one of the two choices below. Which would you rather choose? Reward for detecting an intruder: You have defeated an opponent's attacks that you estimate has cost the opponent 4,000 dirhams. You have two choices as a potential reward: A: 25% chance to get 24000 dirhams; 75% chance to get 0 B: 25% chance to get 16000 dirhams; 25% chance to get 8000; 50% chance to get 0
(A14) You hack into an opponent's account, and you know you only have a limited time before being detected. If you are detected, you may have to pay a certain fine, according to the following two choices: A: 25% chance to pay 24000 dirhams; 75% chance for no fine B: 25% chance to pay 16000 dirhams; 25% chance to pay 8000; 50% chance for no fine (D14) The attacker may be able to breakthrough your firewall or other defense with the potential losses to you. You have two choices, depending on the strength of your defense: A: 25% chance to lose 24000 dirhams; 75% chance for no loss B: 25% chance to lose 16000 dirhams; 25% chance to lose 8000; 50% chance for 0 loss
(A15) You hack into an opponent's account, and you know you only have a limited time before being detected. You may gain some resources according to the following two choices: (D15) Reward for detecting an intruder: You have defeated an opponent's attacks that you estimate has cost the opponent 4,000 dirhams. You have two choices as a potential reward: A: 0.1% chance to get 20000 dirhams; 99.9% chance to get nothing B: 100% chance to get 20 dirhams
(A16) You hack into an opponent's account, and you know you only have a limited time before being detected. If you are detected, you may have to pay a certain fine, according to the following two choices: (D16) The attacker may be able to breakthrough your firewall or other defense with the potential losses to you. You have two choices, depending on the strength of your defense: A: 0.1% chance to pay 20000 dirhams; 99.9% chance for no fine B: 100% chance to pay 20 dirhams

5 Results

The results presented below are based on sample sizes of responses that are the same in the cases of the Cyberattackers (n = 71 for all questions) and the Cyberdefenders (n = 67 for all questions). In the case of the KT study, the number of respondents varied throughout the questions from 64 to 141 but with an average of 78.8.

The following table demonstrates the results of the 16 test questions, presented as follows. The 16 questions are enumerated in the left-hand column; subsequently the next 3 columns indicate first the percentage of responses from "Cyberattackers" on the left-hand side of the question text, then the percentage of responses from the right hand side of the question text, and finally whether or not the majority of responses agree (Yes) or disagree (No) with the corresponding KT responses (Table 3).

Table 3. Results of Current Testing Compared to KT Results.

	A		AKT	D		AKT	KT	
Q	L%	R%	Y/N	L%	R%	Y/N	L%	R%
1	44	59	Y	51	49	N	18	82
2	44	56	N	48	52	N	83	17
3	18	82	Y	34	66	Y	20	80
4	45	55	N	40	60	N	65	35
5	42	58	Y	54	46	N	22	78
6	41	59	N	46	54	N	67	33
7	32	68	Y	48	52	Y	14	86
8	61	39	Y	40	60	N	73	27
9	45	55	Y	34	66	Y	20	80
10	27	73	Y	84	16	N	22	78
11	44	56	Y	57	43	N	16	84
12	63	37	Y	66	34	Y	69	31
13	11	89	Y	16	84	Y	18	82
14	44	56	N	48	52	N	70	30
15	66	34	Y	57	43	Y	72	28
16	43	57	Y	51	49	N	17	83

Legend:
Q = Question
D = Cyberdefender Responses
L = "Left-hand Choice" ("A" or "C")
KT = Kahnemann/Tversky Responses
R = "Right-hand Choice" ("B" or "D")
AKT = Agreement with KT
A = Cyberattacker Responses

As a first-level analysis, we would conclude current results in prospect theory do not necessarily hold when reinterpreted in the cybersecurity context. However, we ignored Q3 because its phrasing could lead to a contradictory conclusion when compared to KT. In this analysis, we indicate agreement or disagreement with the KT results if the majority response is the same in each case. We note that the Cyberattacker responses only coincide with KT responses 73% of the time; and even less of an agreement between the Cyberdefender responses and KT, namely 33% of the time.

It should also be noted that there is a distinction in terms of the presentation of the test questions. In questions 1 through 8, although the subjects are asked to assume that they are either a Cyberattacker for a Cyberdefender, the nature of the questions does not assume that the monetary values defined in the question need to be interpreted as specific values in their cyber environment, such as the cost of memory, the cost in time of a program execution, or the cost of other machine components. In questions 9 through 16 for the Cyberattackers or Cyberdefenders, the monetary costs are more directly aligned to their machine operations.

When splitting the responses into the agreement or disagreement with KT, we describe the responses to questions 1–8 as "first half", and the latter as "second half". In the first half, Cyberattackers agree with KT 57.1% of the time, and 87.5% in the second half. Cyberdefenders agree with KT 28.6% of the time, and 50.0% in the second half.

6 KT Challenges to Prospect Theory

Kahneman and Tversky demonstrated that classical utility theory did not hold when human factors such as perceived risk were tested by posing questions that would lead to the same choices by a rational actor, but in many test cases respondents would choose opposite values when utility theory would predict they would choose the same. In numerous examples, they showed that often, options that should lead to the same conclusions may have vastly different results, thus violating the classical theory of utility, as indicated below.

The purpose in our research was to see if the contradictions to classical utility theory would also be true when posed in a cybersecurity environment (Table 4).

Table 4. Distinction Between Current Results and KT Results for Selected Pairs of Questions.

Problem pair	Level of agreement	Attacker 1st v 2nd	Defender 1st v 2nd	KT 1st v 2nd
Problem 1/Problem 2	Neither	0	−3%	+65%
Problem 5/Problem 6	Neither	−1%	−8%	+45%
Problem 7/Problem 8	Attackers	+29%	−8%	+45%
Problem 9/Problem 10	Defenders	−18%	+9%	+59%
Problem 11/Problem 12	Both	+19%	+9%	+2%
Problem 13/Problem 14	Both	+33%	+32%	+52%
Problem 15/Problem 16	Both	−23%	−6%	−55%

7 Conclusions

The analysis of the KT results compared to the similar analysis of comparable questions when posed in the context of the cybersecurity environment may lead one to believe that the interpretation of the relative importance of economic values in the cyber context may be different than in a population as a whole. There are a number of potential explanations for this difference. It is conceivable, for example, that many persons interested in cybersecurity do not focus on the economic value or cost of cyber activities. It may also be that the challenge of trying to assess what is familiarly thought of as memory space, machine cycles, software, communications or equipment costs are difficult to assign to normal computer for network usage.

References

Kahneman, D., Tversky, A.: Prospect theory: an analysis of decision under risk. Econometrica **47**(2), 263–291 (1979). https://doi.org/10.2307/1914185.JSTOR1914185

Patterson, W., Winston-Proctor, C.: Behavioral Cybersecurity. CRCPress, Orlando (2019)

Representing a Human-Centric Cyberspace

Phoebe M. Asquith[1,2] and Phillip L. Morgan[1,2(✉)]

[1] Airbus, The Quadrant, Celtic Springs Business Park, Newport NP10 8FZ, UK
{phoebe.p.asquith.external,phillip.morgan.external}@airbus.com,
morganphil@cardiff.ac.uk

[2] School of Psychology, Cardiff University, 70 Park Place, Cardiff CF10 3AT, UK

Abstract. There is a lack of consensus when using the term "cyberspace" [1]. Computers and network devices are prominent in definitions of cyberspace; less common is the essential and inclusion of human users. However, the human user is both implicitly integral to and actively part of the cyberspace.

Cyberspace is often conceptualized as three layers of interconnected networks: social, information and geospatial (physical) [2]. These represent an indirect human element within cyberspace. This is characteristic of related fields, such as cybersecurity, where human-centered research has been lagging behind technological aspects. A model that incorporates the human user in cyberspace is needed to direct future research and improve security and usability (navigation).

A new human-centric model of cyberspace is proposed (the HCCM), with the user as a physical and integral entity, together with recognition of the cognitive representation of cyberspace. It focuses on boundaries and transformation points between objects and spaces and offers a platform for future human-centric research in cybersecurity.

Keywords: Cybersecurity · Cyber security · Human user · Human factors · Human-machine interaction

1 Introduction

Metaphors and analogies, such as "wild west" and "space", have been central to attempts to understand the global online computer network and its meaning for society and culture broadly [3]. The term "cyberspace" was first used by William Gibson in his book, Neuromancer [4], where he defined it as a "a consensual hallucination". Since then, although the term "cyberspace" is commonly used, there is a lack of consensus about its meaning and what it encapsulates [1, 5]. It is difficult to represent and model cyberspace, due to its associations across physical (e.g. computer hardware) and non-physical domains (e.g., 'information' or 'online' space).

Although computers and network devices are prominent in current common definitions, less common is the inclusion of human users. Cyberspace has been described within

I. Corradini et al. (Eds.): AHFE 2020, AISC 1219, pp. 122–128, 2020.
https://doi.org/10.1007/978-3-030-52581-1_16

dictionary[1] and literary sources with themes such as "communication", "virtual", "electronic", "network" and "computer" [5–7]. Kautz [8] identifies hardware and software as "universals" within cyberspace, and computers and computer networks as preconditions for cyberspace. We strongly assert that humans (e.g., system/computer/network users) are also preconditions for cyberspace: the human user is both implicitly integral to the creation of cyberspace and actively part of the cyberspace.

The presence of human-human connections and "communities" within the virtual space is widely acknowledged in research literature and common understanding [9]. Popular social media sites have billions of active users across the globe sharing information (for example Facebook, which had ~2.45 billion monthly active users in 2019 [10]). Information transferred from human cognitive and physical space into the digital realm has been conceptualized as an extension of the self within the virtual space [11–13]. Implications of human actions within cyberspace in law are becoming more widely discussed [14], and language (a human faculty) is mapped in cyberspace [15]. Models of risk within cybersecurity also include the interaction between humans and technology [16]. Despite this, the human user has not often been included within dynamic models of cyberspace. Crucially, a clearer model of the importance of the human user in cyberspace is needed, to direct future research and improve the security and usability (navigation) of cyberspace. A key aim of the current paper is to provide the structure of such a model, that we aim to develop further.

The "cyberspace" concept relies on a cognitive representation of a "space" within which information is shared. Human interactions with online devices provide a "window" into this virtual space. If humans are not integral to cyberspace:

a) The technology itself would not exist;
b) [Putting aside the human creation and development] All systems would need to be fully autonomous;
c) Large aspects of data movement would be neglected;
d) Signals would be reduced to binary connections with no "space" conceived.

Cyberspace has traditionally been modelled as having three layers of interconnected networks; geospace (physical), infospace (logical and virtual) and sociospace (social and political) [1, 2]. A cyber-physical system (CPS) facilitates the information communication across the three layers. The movement of information ultimately affects outcomes and decisions at sociospatial endpoints. For example, one model, based on cybernetics – "the scientific study of control and communication in the human and the machine" by Norbert Wiener [17, 18], identifies engineering [*geospace*] and software [*infospace*] aspects of machines in cyberspace, and, economic and socio-cultural aspects [*sociospace*] relating to human users [19]. These models represent, to some extent, a human end-point element to cyberspace, however it is an indirect role. Using models of cyberspace with individual and connected human users directly incorporated is crucial to help to guide research in human factors and socio-cultural aspects of cybersecurity, and

---

[1] *Lexico dictionary* [32]: "The notional environment in which communication over computer networks occurs". *Collin's English Dictionary* [36]: "In computer technology, cyberspace refers to data banks and networks, considered as a place".

beyond. For example, to embed improved cybersecurity practice within a staff culture, a multi-faceted and human-centric approach is needed [20–22].

Hai Zhuge has, for the past 10 years or so, been exploring the concept of a social-cyber-physical space and the future of a connected cyber-human world [23]. In a world with seamless cyber-human relationship, Zhuge envisaged the coming together of the Physical, Virtual and Mental [24–26]. The inclusion of 'Mental' here begins to allude to a vision of a cyberspace that can includes a cognitive element, to better direct human factors and psychological research in human-machine interaction.

The importance of human cognition in cyber-human interactions is also posited by Jinhua and colleagues [27], who describe cyberspace as a space parallel to traditional space, into which we project "simulations" of physical and social elements [28–30]. This is a useful concept, as it acknowledges humans as the creators of this projection, and places them as prerequisites to cyberspace. This is also key when directing research in the usability of systems and architecture.

A network model created by Hao et al. [31] places humans as integral to activity within cyberspace and aims to help with analyzing threats in the cyberspace. The inclusion of humans is crucial to this process and with improving defence within cybersecurity; human error plays a major role in cyber-breaches [32]. This model begins to explore the importance of humans within the cyberspace but does not fully address the interaction between humans and the physical or information space.

Each of these models has made important developments in conceptualizing a cyberspace that helps to guide research in the area. A new model of cyberspace is presented below, that brings together some of these ideas, to provide a more human-centric model. A network representation from the physical, cognitive and social human user, to the physical device and virtual spaces will help guide research in fields such as cybersecurity

2 A New Model of Cyberspace

Our new Human-Centric Cyberspace Model (HCCM) includes various novel concepts:

- The user as a physical and cognitive entity;
- The user as integral within cyberspace;
- The cognitive representation of cyberspace;
- A focus on boundaries and transformation of information between elements.

The HCCM identifies humans, devices and systems as key objects within cyberspace. Information is transferred between humans and systems through devices that have hardware and software elements. Humans and systems are connected through a network of activity, which is dynamic and driven by the goals of human users and architects. Within HCCM, "cyberspace" is the cognitive representation of this activity space.

Human users exist in the physical space, interacting with physical devices. The physical human applies perceptual (e.g., seeing information, attending to information) and cognitive (e.g., decision-making, judgment, reasoning) abilities that process the information presented via a device (e.g., the human-machine interface/HMI). "Social" or "cultural" aspects are also important manifestations within cyberspace: they are projections and representations created by human users, which are reliant on cognitive and perceptual processes (see Fig. 1). Data transferred between systems is within the "infospatial" (not physical) space. The connected infospace is reliant on software, which in turn is reliant on hardware for function (see Fig. 1).

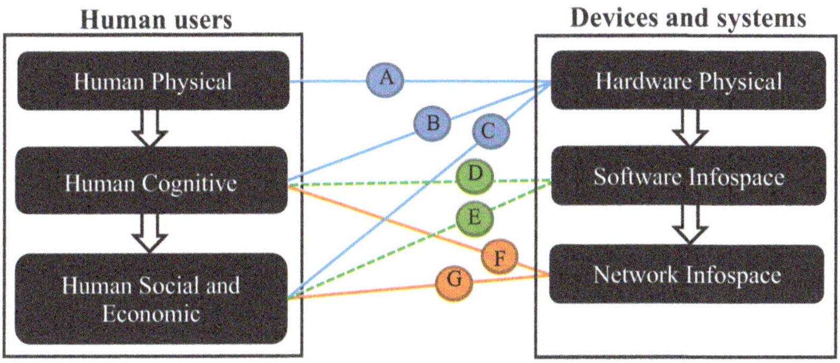

Fig. 1. A human centered cyberspace model, to be used as a high-level tool to guide cyber research: see Table 1 for suggested research areas based on connections.

Each of the levels in the human and machine aspects of cyberspace have important interactions across the human-system/machine boundaries (see Fig. 1). Mapping these connections and boundaries can guide research concerning humans within cyberspace such as human-machine interaction and human-centered cybersecurity (see Table 1).

Each of the connections indicates an area of research that can be further broken down into research modules, inspiring topics and human factors guided research; an area that has been lagging behind a technological focus in areas such as cybersecurity. This model can aid with preventative management of cyber threats within a human-machine connected cyberspace by turning the risk of human error into an opportunity for research and improvement and by crucially including human-users in the planned solution [33].

Table 1. Research areas identified using connections between elements of the HCCS.

	Description/Key questions	Example research areas
A	Human users interacting with physical devices. How does design the of devices affect human usage?	Human-machine interaction; Human factors; Usability of hardware
B	Humans processing information received on physical devices e.g. visualization techniques, nudges. How can hardware support decision-making and improve efficiency?	Human-machine interaction; Human factors; Usability; Decision-making
C	Hardware availability across different cohorts. How does this technology support economic and social projections? How do technologies interact with multiple users across different cultures?	Inclusion; Technological advances
D	Human interaction with software technologies. How does software support cognitive processes to achieve goals?	Cognition; Decision making
E	Software availability across different cohorts. How does software support economic and social projections? How does interact with multiple users across different cultures and support shared goals?	Inclusion; Technological advances
F	How does network architecture support human use and decision-making? How do humans represent a connected information space?	Cognition; Human-machine teaming; Decision- making
G	How do connected networks support economic and social projections? How does a connected network interact with multiple users across different cultures and support shared goals?	Digital communications; Socio-economics; Policy and borders

3 Conclusion

Within the current paper, we highlight the lack of research and associated literature that models and considers humans as integral within cyberspace. Drawing upon some recent examples where humans have been considered to some extent [2, 19, 25, 27, 31] *and* through consideration of human physicality, cognitive abilities (e.g., perception, attention, thinking) and human social and economic factors, we have developed and present the first version of the HCCM. With humans included within conceptualizations of cyberspace, the model allows for important considerations to be recognized as areas for research investigation within the field of human-centric cybersecurity, and beyond. Next steps will involve further developing this high-level model to a more specific cyberspace concept and research guide.

Acknowledgments. The research and Airbus Accelerator in Human-Centric Cyber Security (H2CS) is further supported by Endeavr Wales and Cardiff University. The second author (Dr Phillip Morgan) as Technical Lead, and the first author (Dr Phoebe Asquith) as a Cardiff University Research Associate working at Airbus.

References

1. Kademi, A.M.A., Koltuksuz, A.: Formal characterization of cyberspace for cyber lexicon development. In: ECCWS 2017 16th European Conference on Cyber Warfare and Security, p. 200 (2017)
2. Bayne, J.: Cyberspatial mechanics. IEEE Trans. Syst. Man Cybern. Part B **38**(3), 629–644 (2008)
3. Johnston, R.: Salvation or destruction: metaphors of the Internet. First Monday **14**(4) (2009)
4. Gibson, W.: Neuromancer. Berkley Publishing group, New York (1989)
5. Madnick, S.E., Choucri, N., Camiña, S., Woon, W.L.: Tow (ards better understanding Cybersecurity: or are "Cyberspace" and "Cyber Space" the same? mit.edu (2014)
6. Mayer, M., Martino, L., Mazurier, P., Tzvetkova, G.: How would you define Cyberspace, First Draft Pisa, **19** (2014)
7. Kramer, F.D., Starr, S.H., Wentz, L.K.: Cyberpower and national security, Potomac Books Inc. Nebraska (2009)
8. Kautz (2010). https://www.slideshare.net/study4cyberwar/cyberspace-model. Accessed 1 Nov 2019
9. Rheingold, H.: The Virtual Community: Finding Connection in a Computerized World. Addison-Wesley Longman Publishing Co., Inc., Boston (1993)
10. Clement, J.: Facebook: number of monthly active users worldwide 2008–2019, Statista, 19 November 2019. https://www.statista.com/statistics/264810/number-of-monthly-active-facebook-users-worldwide/. Accessed 7 Jan 2020
11. McLuhan, M.: Understanding Media: The Extensions of Man. MIT Press, Cambridge (1994)
12. Zylinska, J.: The Cyborg Experiments: The Extensions of the Body in the Media Age. A & C Black, London (2002)
13. Clark, A., Chalmers, D.J.: The extended mind. Analysis **58**, 10–23 (1998)
14. Lessig, L.: The zones of cyberspace. Stanf. Law Rev. **48**, 1403 (1995)
15. Ivkovic, D., Lotherington, H.: Multilingualism in cyberspace: conceptualising the virtual linguistic landscape. Int. J. Multilingualism **6**(1), 17–36 (2009)
16. Leuprecht, C., Skillicorn, D.B., Tait, V.E.: Beyond the Castle Model of cyber-risk and cyber-security. Gov. Inf. Quart. **33**(2), 250–257 (2016)
17. Wiener, N.: Cybernetics or Control and Communication in the Animal and the Machine. MIT Press, Cambridge (2019)
18. Duffy, P.R.: Cybernetics. J. Bus. Commun. **21**(1), 33–41 (1984)
19. Vinnakota, T.: A cybernetics paradigms framework for cyberspace: key lens to cybersecurity. In: IEEE International Conference on Computational Intelligence and Cybernetics, pp. 85–91 (2013)
20. Bada, M., Sasse, A.M., Nurse, J.R.C.: Cyber security awareness campaigns: why do they fail to change behaviour? arXiv preprint arXiv:1901.02672 (2019)

21. Rayne Reid. V.N.J.: From information security to cyber security cultures. Inf. Secur. South Afr. IEEE 1–7 (2014)
22. Tonge, A.M.K.S.S.C.S.R.: Cyber security: challenges for society-literature review. IOSR J. Comput. Eng. **2**(12), 67–75 (2013)
23. Zhuge, H.: Multi-Dimensional Summarization in Cyber-Physical Society. Morgan Kaufman, Burlingtonn (2016)
24. Zhuge, H., Shi, X.: Toward the eco-grid: a harmoniously evolved interconnection environment. Commun. ACM **47**(9), 78–83 (2004)
25. Zhuge, H.: Future interconnection environment. Computer **38**(4), 27–33 (2005)
26. Zhuge, H.: Future interconnection environment–dream, principle, challenge and practice. In: International Conference on Web-Age Information Management, Berlin, Heidelberg: Springer (2004)
27. Jinhua, L., Xu, H., Zhou, X., Lin, F., An, J.: Research of cyberspace architecture. In: International Conference on Cyberspace Technology (CCT 2013), pp. 367–369 (2013)
28. Zhuge, H.: arXiv (2018). https://arxiv.org/abs/1805.00434. Accessed 12 Dec 2019
29. Miller, R.A., Kuehl, D.T.: Cyberspace and the "First Battle" in 21st-century War. Defense Horizons **68**(1), 1–6 (2009)
30. Czosseck, C., Geers, K.: Borders in cyberspace: can sovereignty adapt to the challenges of cyber security. In: The Virtual Battlefield: Perspectives on Cyber Warfare, vol. 3, no. 88 (2009)
31. Hao, Y., Guo, S., Zhao, H., Chen, Z.: Study on the modeling and analyzing of the role-based threats in the cyberspace. In: IEEE 2nd International Conference on Cloud Computing and Intelligence Systems **3**, 1302–1306 (2012)
32. Security, I.: ZXForce Threat Intelligence Index. IBM Corporation, Armonk (2019)
33. Tàbara D., Saurí, D., Cerdan, R.: Forest fire risk management and public participation in changing socioenvironmental conditions: a case study in a Mediterranean region. Risk Anal. Int. J. **23**(2), 249–260 (2003)
34. Lexico. https://www.lexico.com. Accessed 3 Dec 2019
35. Techopedia Dictionary, https://www.techopedia.com/dictionary. Accessed 3 Dec 2019
36. Merriam-Webster Dictionary. https://www.merriam-webster.com. Accessed 3 Dec 2019
37. Oxford Learner's Dictionary. https://www.oxfordlearnersdictionaries.com. Accessed 3 Dec 2019
38. "Collin's English Dictionary. https://www.collinsdictionary.com/dictionary/english. Accessed 3 Dec 2019

Trust in News and Information in Social Media

Abbas Moallem(✉)

UX Experts, LLC, Cupertino, CA, USA
Abbas@uxexperts.com

Abstract. Social media have significantly changed news consumption. The social media enables users to share anything through a single click on the screen and makes it easy to spread different kinds of information from the news article to the social network. Little effort is made to validate the authenticity of the information and how the social circle is influenced by such news. In this study, we attempt to analyze how much people get the News from social media and if they check the authenticity of the news.

The study includes two parts: a pilot study to observe the people's actual behavior with news posting on social media and a survey.

The survey is designed to collect quantitative data with an online survey. The survey has been administered to students from a public university in Silicon Valley, California, in 2020. This paper summarizes the finding and analysis of the results.

Keywords: Social media · Cybersecurity · Human-computer interaction · News

1 Introduction

Mobile phones, computers, and laptops are tools that provide the ability to explore information across the globe and communicate with people through texts, images, and videos. People use social media all the time at home and in their workspace to share their thoughts and joyous moments. Not only can one communicate with their friends, but people can also sneak peeks into a stranger's profile anonymously. According to Pew Research, Facebook and YouTube dominate social Media, as notable majorities of U.S. adults use each of these sites. At the same time, younger Americans (especially those ages 18 to 24) stand out for embracing a variety of platforms and using each of them frequently. Some 78% of 18- to 24-year-olds use Snapchat, and a sizeable majority of these users (71%) visit the platform multiple times per day. Similarly, 71% of Americans in this age group now use Instagram, and close to half (45%) are Twitter users [1].

People rely on social media to preserve relationships. People also use social media to these days to get information and read the news. Studies suggest that social media also influence collective action and helps in mobilizing protests [2].

Unfortunately, such social media channels open the gateway for cybersecurity criminals and influencers for exploitative purposes, often circulating fake news and information to influence the opinions of a population. People are lured to read what appears to be an important news article or major information, and instead end up clicking on

I. Corradini et al. (Eds.): AHFE 2020, AISC 1219, pp. 129–134, 2020.
https://doi.org/10.1007/978-3-030-52581-1_17

links that act as a nefarious cybersecurity danger. This occurs because those links lead to websites that run malware which is able to extract sensitive information from the user. The website may force the user to download the executable and inject the virus into the user's system. Users should be careful about clicking on any suspicious links.

Eye-catching headlines are frequently used in social media to encourage users to enter a malicious website or download malware, as well as circulate such malicious websites. Similarly, when those headlines state false or misleading information, their circulation on social media is massively used to influence people's opinions by propagating such fake news. The most prominent case was how Facebook was influencing and shaping the voting behavior people in the 2016 United States election [3]. After the election, there was a widespread revelation as to what most experts already knew about social networking personal data collection. The Cambridge Analytica data scandal is now common knowledge, with most people understanding that the personal data of millions of peoples' Facebook profiles was harvested without consent and then used for political advertising purposes. However, before the U.S. elections, nobody questioned what Cambridge Analytica were doing in other countries. Cambridge Analytica has now ceased operations, but what happens with the tools that they built and the data that they collected remains unknown. It is possible that both the tools and data are still used by other unknown entities [4].

Fake news does not just influence elections and political perceptions; misinformation and fake information also indirectly shapes population behaviors on scientific issues such as medical and pharmaceutical matters [5–9].

The objective of this study is to measure the amount of trust placed in news information from social media among the younger population in the Bay Area of California, USA, one of the most technologically and economically advanced regions and ethnically diverse populations.

2 Method

The study includes two parts: a survey and a pilot study to observe people's actual behavior with news postings on social media.

The survey was designed to collect quantitative data through an online survey. The survey administered to students from a public university in Silicon Valley, California in 2020. The survey was administered to students enrolled in engineering courses.

The survey includes the following ten questions:

- How frequently do you listen, watch or read News using any available channels?
- What kind of News do you generally follow?
- Approximately what percentage of your news/information is coming from social media?
- How often do you check the News from social media?
- Which social media platform do you receive most News from?
- How trustworthy do you find news/information reports on social media? (One Least trustworthy, ten most trustworthy): Facebook, Twitter, Instagram, Reddit, YouTube, Tumblr, Pinterest, LinkedIn, Google+, Other 1 and Other

- Your gender: Male, Female and Other
- Your Age: Under 25, 25–34, 35–44, 45–54, 55–64 and over 65
- Status: Engineering-Undergraduate, Engineering Graduate, Undergraduate: all other disciplines and Graduate: All other disciplines

To date, 89 students have completed the survey. We did not collect any personal identifying data about respondents. 46% of respondents are female, and 54% are males. 78% are 21–30 years old, 8% are under 21, 11% are 30–40 years old, and 2% are over 40 years old. Thus, overall the respondents are young, as expected for college students. Around 48% are undergraduates (38% engineering students and 10% other disciplines) and 52% are graduate students (32% in engineering graduate and 22%, other disciplines.

A pilot study was also conducted by students enrolled in my cybersecurity course to investigate the actual behavior of users who encounter a news article posted on social media, and, without checking the authenticity of the news, click the posting on social media. The purpose was also to increase awareness of the participants about the risks on social media.

The pilot study was designed to investigate whether people are more likely to read fake news posted on social networking sites like Facebook, and whether people are more likely to read the fake news if the title is attractive or if the news is related to some social activity.

For the purposes of the study, a few fake posts on different social media websites were created and posted on the three most popular social networking websites: Facebook, Twitter, and LinkedIn. Each of these sites also allow any user to post content easily without being subject to any kind of real security checks. Other users can like, comment, or share these posts.

- On Facebook, the post was titled "See latest new Game of Thrones Trailer" and an embedded picture of "Night king of Game of Thrones." Upon clicking of this post, the user is redirected to fake blogging websites that were created for this experiment instead of any official Game of Thrones websites (Fig. 1).
- On LinkedIn, the post was titled "Google Launched New Gaming Platform, Claim Free Access Today" and attached a picture of the Google logo to make it appear authentic. Upon clicking this link, the user is redirected to a fake gaming website that was created for this experiment where visitors able to see ads for many unregistered online games (Fig. 2).
- On Twitter, a tweet of fake exit polls for the 2020 election of the USA and included an embedded picture of President Trump and a link. Upon clicking the link, the user is redirected to fake news website, created for this experiment (Fig. 3).
- The posts were monitored for a few days and data was gathered about the number of post shares, number of comments, and number of likes. Then, a survey link was sent to those who clicked or shared the post, explaining that this was a social experiment and we were trying to spread awareness about fake news. We asked if they would answer the survey. Fifty people responded to the survey.

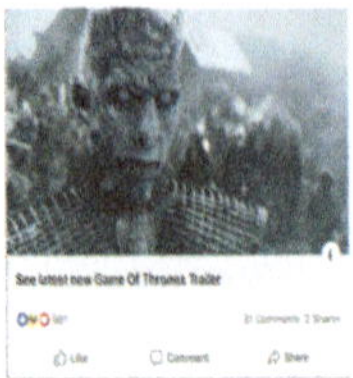

Fig. 1. On Facebook, the post was titled "See the latest new Game of Thrones Trailer."

Fig. 2. On LinkedIn, the post was titled "Google Launched New Gaming Platform, Claim Free Access Today.

Fig. 3. On Twitter, a tweet of fake exit polls of the 2020 election of the USA and embedded picture of President Trump was created for this experiment

3 Results

Pilot Study

Almost all participants (47/50), who completed the survey after having clicked on one of the fake posts, declared that they use social networking sites like Facebook, Instagram, Twitter, etc.

The survey asked which forms of social media these participants used most often. The options provided to the users in the survey included Facebook, LinkedIn, Instagram, Twitter, Google Plus, among others. According to this survey, we found that most of the people use LinkedIn, followed by Instagram, Twitter, and Facebook. This tells us that people these days are more dependent on these social networking websites to get information about news, sports, education, entertainment and so on.

Then we asked the users which news websites they use the most. Users were provided answer options including CNN, BBC, Fox News, MSNBC, and others. The response to this question was positive in the case of two websites: CNN and BBC. Fox News was the third most selected option. This shows that people rely on popular news websites to gather information. If these news channels start to show fake news or if the websites of such news channels are hacked by the attacker, they can gain access to important information and can share fake news using these channels as their medium. Since people trust these news channels more, they might believe in this fake News.

Survey takers were also asked how they prefer to gain knowledge to get information about what's happening around. We asked whether they prefer social media or news channels. The survey results showed that people are more likely to use social media, and about 80% (of the fifty surveyed people) prefer to use social media over news channels.

The participants were asked whether they have seen any fake news on the social networking sites. If so, we asked which site they had seen the fake news on. 31% answered Twitter as the social networking site where they have seen fake news and information, while 27% said Facebook, and 13% said Instagram.

Survey Results

How frequently do you listen, watch or read news using any available channels?

The results of the survey indicate that 43% of the participants read news every day, 19% once in a while, 14% multiple times during the day, 20% every week, and 6% never.

What kind of news do you generally follow?

31% of participants generally follow technology news, 18% follow domestic politics, 21% follow international politics, 14% follow arts and entertainment, 9% follow sports, and 8% follow other types of news.

Approximately what percentage of your news/information is coming from social media?

29% of participants claim that 30 to 60% of their News and information is coming from social media, 24% of participants say 60 to 80%, 16% of participants 10% to 30%, 21% of participants say over 80% and 10% of participants just 10%.

How often do you check news from social media?

61% check social media every day, 12% every other day, 2% more than once per week, 21% once in a while, and 2% more than once per week.

Which social media platform do you receive most News from?

19% receive most news from Facebook, 13% Twitter, 17% Instagram, 21% YouTube, 12% LinkedIn, 17% other social medias and only 1% said they don't get News from social media.

How trustworthy do you find news/information reports on social media?

The results indicate that college students do rely on social media to get news. According to this survey, 38% use Facebook to get the news and 22% use Twitter, which might be considered as a more reliable source.

4 Conclusion

The results of this survey indicate the overall participants rely on social media to get news. The sharing of news media on social networking sites might influence participants considering the sources of news on social media might not be trustworthy. The result of the survey indicates that college students actively share information without knowing more about the news source, because it is easy and takes no effort. The results of this study indicate how many options of the people might be shaped by what is known as "Fake News."

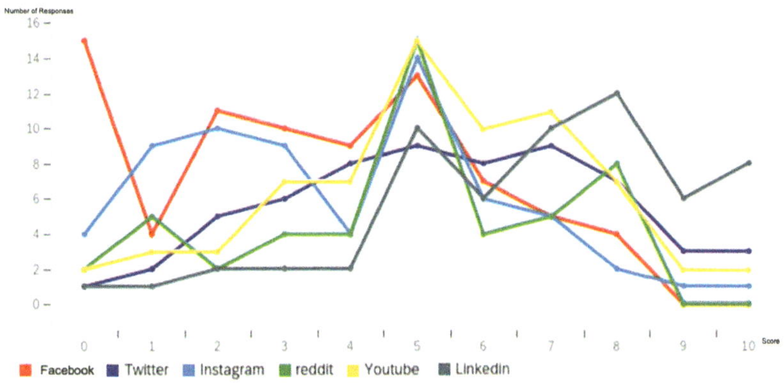

Chart 1. Social networking sites' trustworthiness pattern as rated by 89 participants (1- least trustworthy, 10- most trustworthy).

The score pattern of the trustworthiness of news illustrate that even though participant trust 5 out 10 points, however the participants don't highly trust the news on social media (Chart 1).

This study show showing the rend further qualitative and quantitative studies is needed to validate further this study.

References

1. Smith, A., Anderson, M.: Social media use in 2018. Pew Research Center, 1 March 2018. https://www.pewresearch.org/internet/2018/03/01/social-media-use-in-2018/
2. Jihyang, C., et al.: Investigating effects of social media news sharing on the relationship between network heterogeneity and political participation. Comput. Hum. Behav. **75**, 25–31 (2017)
3. Allcott, H., Gentzkow, M.: Social media and fake news in the 2016 election. J. Econ. Perspect. **31**(2), 211–236 (2017). https://doi.org/10.1257/jep.31.2.211
4. Cadwalladr, C., Graham-Harrison, E.: Revealed: 50 million Facebook profiles harvested for Cambridge Analytica in major data breach, The Guardian, 17 March 2017. https://www.theguardian.com/news/2018/mar/17/cambridge-analytica-facebook-influence-us-election
5. Waldrop, M.: News feature: the genuine problem of fake news. PNAS **114**(48), 12631–12634 (2017). https://www.pnas.org/content/114/48/12631
6. Vosoughi, S., Roy, D., Aral, S.: The spread of true and false news online (2019)
7. Tim, D., Fiona, M.: Sharing news online: social media news analytics and their implications for media pluralism policies. Digital J. **5**(8), 1080–1100 (2017)
8. Fiegerman, S.: Influencing and shaping vetoing behavior of the people through Facebook on the USA election in 2016, CNN, 17 November 2016. https://money.cnn.com/2016/11/17/technology/facebook-election-influence/
9. Wylie, Ch.: How Trump Consultants Exploited the Facebook Data of Millions. New York Times, 17 March 2018. https://www.nytimes.com/2018/03/17/us/politics/cambridge-analytica-trump-campaign.html

Is Data Protection a Relevant Indicator for Measuring Corporate Reputation?

Isabella Corradini[1,3](✉) and Enrico Nardelli[2,3]

[1] Themis Research Centre, Rome, Italy
isabellacorradini@themiscrime.com
[2] Department of Mathematics, University Roma Tor Vergata, Rome, Italy
nardelli@mat.uniroma2.it
[3] Link&Think Research Lab, Rome, Italy

Abstract. Over the last few years the importance of reputation has grown both for individuals and organizations, especially because of the Internet and social media platforms. Considering the value of data and information, corporate reputation also passes through companies' ability to protect sensitive customers' data. When compromised, after a cyberattack or a data breach, one of the most important risks for a company is the loss of customers' trust and the negative impact for future business. Therefore, privacy and security data should be considered as a priority for organizations to safeguard trust and business. In literature, models measuring reputation consider several dimensions, such as leadership, vision, corporate social responsibility, emotional attractiveness. In this paper we analyse the relationship between cyber-threats and reputation and, on the basis of models available in literature, we discuss the possibility of including data protection among indicators for measuring corporate reputation.

Keywords: Reputation · Cybersecurity · Trust · Communication

1 Introduction

Over the last few years, because of the Internet and social media, the importance of reputation has grown. Internet represents a formidable source of information, able to affect people's impressions; moreover, social media platforms, developed to connect people, are more and more used by organizations for their online business, and increasing marketing in a different way [1].

Reputation is a multidimensional construct which includes different meanings [2], and that can be interpreted under several perspectives [3], such as economics, sociology, organizational behaviour. Besides individuals, reputation regards also organizations (corporate reputation), generated by the estimation of internal and external stakeholders.

The perceptual element plays a fundamental role, since corporate reputation derives from an amalgamation of perceptions and opinions developed by its different stakeholder [4], and reflects people's perception [5].

I. Corradini et al. (Eds.): AHFE 2020, AISC 1219, pp. 135–140, 2020.
https://doi.org/10.1007/978-3-030-52581-1_18

Considering the digital era and the value of data, and given that trust contributes to the building of corporate reputation, it is obvious the relevance for companies to handle cybersecurity effectively, in order to ensure the protection of customers' data. Indeed, a company which fails to protect them might compromise trust and generate a risk for future business. Because data breaches provide a high risk to company's reputation, data privacy and security are perceived by leaders as a priority to manage [6].

In this paper we discuss the concept of reputation and its relationship with cybersecurity, and the issue of including data protection among the indicators usually applied for measuring reputation, according to the available models in literature.

2 About the Concept of Reputation

Research about corporate reputation has been developed especially in management research [7], as shown by the amount of publications in this area [8].

It is a shared opinion that corporate reputation represents a multidimensional and dynamic construct [3, 9, 10]. The main models for the measurement of reputation have focused on the plurality of these dimensions, according to different scholars' orientations. However, reputation represents a social construction [11], where communication exchanges, whatever in physical or digital world, play an important role.

In our view [12], perceptual elements and social relationship with the various stakeholders are the prevailing elements that define reputation. In this sense, corporate reputation can be defined as [13] «*the result of the interactions between a company and its stakeholders, which include customers, users, suppliers, internal staff, consultants, etc. All these actors form impressions and develop evaluations about company activities, so directing their own behavior and affecting that of others*».

In the digital era further factors have to be considered for construction and maintaining of reputation. Given the wide opportunity for companies to conduct business on the Internet, the issues of trust and trustworthiness play a strategic role [14, 15]. When something happens in the public sphere affecting the perception of the level of trust of a company, this will immediately and directly affect the company reputation.

Two elements characterize the concept of reputation: one temporal and one contextual. The temporal element refers to the fact that reputation is built and consolidated over time; moreover, precisely because it concerns values and perceptions elaborated by stakeholders, it cannot be defined in a static way. Perceptions can change and, consequently, reputation can also take on a different connotation. In the same way, reputation changes according to the context of reference.

The same organization can have a good and bad reputation at the same time, depending on the stakeholders involved. For example, a company that works to protect the environment will elicit a positive feedback by individuals embracing this goal, but not by those who, instead, are interested in exploiting territories for economic purposes.

A good reputation is beneficial and convenient in the long term, since, for example, a company can gain a competitive advantage thanks to the better perception that its stakeholders have of it [16]. Other advantages for companies are greater visibility, protection of their values, and the improvement of their capacity to retain qualified personnel [17]. In a global and uncertain market, the challenge for organizations is to improve their ability to build and maintain customers' confidence.

3 Data Breaches and Reputation

Cybersecurity risks are considered an important operational challenge to be handled by global executives [18], and data security is one of the most relevant key macro-trends for the reputational landscape.

Since cybersecurity is a must and customer's trust is strategic for any business, the connection between cyber-threats and reputation is evident [19]. Indeed, companies which cannot protect consumers' data might compromise their trust, generating a risk for future business. Considering that data breaches are growing and that they are associated with a negative sentiment experienced by customers, companies cannot neglect the issue of data protection if they want to keep a good reputation.

It is plausible that the increasing attention to reputation is partially depending on the current regulations, in particular the EU's General Data Protection Regulation (GDPR). This law has made organizations accountable for the protection of their customers' data, requiring them to consider data protection principles "by design" (i.e., since the development earliest stages) and "by default" (i.e., always processing data with the highest privacy protection level). Any breach affecting rights and freedoms of individuals must then be reported to the relevant supervisory authority, not later than 72 hours.

When a company admits to having suffered from a data breach, this generates doubts about the attention paid to security measures and, generally, to its clients. From a reputational viewpoint, this is a twofold problem: both customers and company are victims of the attack, and both parties suffer from the consequences, even if with different responsibilities. It is therefore evident that, besides offering products and services, an organization should guarantee the best protection for its customers and their data. Moreover, each organization should consider the various types of cost derive from data breaches [20].

Popular cases of data breaches have shown their consequences on company reputation. Just think of the massive security breach that struck in 2015 Ashley Madison, a famous dating website for married people, exposing about 36 million users accounts all over the world, including sensitive information like secret sexual fantasies. Or think of Equifax, one of the most important consumer credit reporting agencies, that in 2017 announced a data breach which exposed the personal information of more than 140 million people. In both cases the problem does not regard the economic aspects only, but the urgent need of restoring consumers' confidence.

4 Data Protection and Reputation Measurement Models

Starting from the multidimensional view of reputation, it is possible to identify the different dimensions that contribute to its measurement. The complexity of this issue suggests that to exhaustively analyze a company's reputation, the best strategy is to follow a multidisciplinary approach, in order to avoid focusing only on those dimensions that do not grasp the real reputation value.

Different criteria can be used to measure reputation [21] depending, for example, on the objective and subjective elements or the nature of stakeholders involved. We consider three main approaches for measuring reputation [12], according to the type of stakeholders involved (generalist or specific) and to the type of evaluation (rational or emotional) of the different dimensions. In this sense, we distinguish three different approaches:

1. The analytical approach addressed to a general audience
2. The analytical approach addressed to a specialist community
3. The synthetic approach.

In the "analytical approach" reputation is evaluated on the basis of the measurements of a variety of cognitive and rational indicators, belonging to many dimensions, which are then weighted and combined in an overall reputation index. The analytic models can be based on a target population of specialists (e.g., financial analysts) or on the general audience. Finally, in the synthetic ones, the basic indicators belong to the sentimental and emotional sphere and are usually less than the ones considered by the analytic approaches.

We report in Table 1 the dimensions and indicators of two well- known models in literature [22, 23] which, according to our classification, are part of the first approach. Given what we have above discussed on reputation, in fact, it is clear that this class is the most affected one by the reputational implications of data breaches.

Table 1. Dimensions and indicators for two corporate reputation measurement models.

Model	Dimensions	Indicators
Reputation quotient (Fombrun et al. 2000)	Emotional appeal	Positive feelings; Admiration and respect; Trust
	Products and services	Support; Innovativeness; Quality; Value for money
	Vision and leadership	Leadership; Vision; Takes advantage of market opportunities
	Workplace environment	Well managed; Good to work for; Have good employees
	Social and environmental responsibility	Support good causes; Environment responsible; Relations with community
	Financial performance	Profitability; Low risk for investors; Outperform competitors; Prospects for future growth
Customer Based Reputation (Walsh, Betty and Shiu 2009)	Customer orientation	Courtesy; Attention; Focus
	Good employer	Good to work for; Treats well people; Leadership

(*continued*)

Table 1. *(continued)*

Model	Dimensions	Indicators
	Financial reliability	Outperform competitors; Recognize market opportunities; Prospect for future growth
	Product and service quality	Reliability; Innovativeness; Support
	Social and environmental responsibility	Efforts to create jobs; Available to reduce profits for a clean environment; Support good causes

In both models, among the indicators measuring the dimensions regarding how well organization is managed, it is not explicitly considered how well the organization protects data of its own customer. However, considering the central role of customers on building and maintaining corporate reputation, we think that data protection should be included as an indicator for measuring reputation.

From our point of view, customers' data protection could be part, for example, of corporate social responsibility [24], or considered anyhow a relevant indicator of customer orientation. Moreover, this could be a sign of more attention towards customer's rights.

Finally, it is important to consider that company's performance in terms of cybersecurity will affect also the general trust of a customer in the organization itself. That is why taking care of cybersecurity has not only an immediate value for any company but also a strategic one.

5 Conclusions and Future Work

In this paper we have discussed the importance of the relationship between cyberthreats and corporate reputation, and the need for companies to handle the protection of customers' data effectively. This is especially true since data security represents one of the most relevant macro-trends for the reputational landscape.

Since reputation management requires an appropriate measurement approach, the issue is how to include data protection among indicators so as to achieve this goal. We think that data protection is an important indicator showing how companies take care of their customers' data, which can favourably affect the development or the maintenance of their reputation.

We recalled the classification of the various reputation measurement models available in literature, depending on the type of evaluation of the different dimensions and the stakeholders involved. We discussed a preliminary proposal on how to extend one class of reputation measurement models so as to include an indicator measuring data protection.

References

1. Ramsaran-Fowdar, R.R., Fowdar, S.: The implications of Facebook marketing for organizations. Contemp. Manag. Res. **9**(1), 73–84 (2013)
2. Barnett, M.L., Jermier, J.M., Lafferty, B.A.: Corporate reputation: the definitional landscape. Corp. Reputation Rev. **9**(1), 26–38 (2006)
3. Fombrun, C.J., Van Riel, C.B.M.: The reputational landscape. Corp. Reputation Rev. **1**(1/2), 5–13 (1997)
4. Bennet, R., Kottasz, R.: Practitioner perceptions of corporate reputation, an empirical investigation. Corp. Commun. Int. J. **5**(4), 224–234 (2000)
5. Rose, C., Thomsen, S.: The impact of corporate reputation on performance: some danish evidence'. Eur. Manag. J. **22**, 201–210 (2004)
6. Reputation Institute: 2020 Global Trends in Reputation (2020)
7. Rindova, V.P., Williamson, I.O., Petkova, A.P.: Reputation as an intangible asset: reflections on theory and methods in two empirical studies of business school reputations. J. Manag. **36**(3), 610–619 (2010)
8. Veh, A., Göbel, M., Vogel, R.: Corporate reputation in management research: a review of the literature and assessment of the concept. Bus. Res. **12**(2), 315–353 (2018)
9. Walsh, G., Beatty, S.E.: Customer-based corporate reputation of a service firm: scale development and validation. J. Acad. Mark. Sci. **35**, 127–143 (2007)
10. Helm, S., Liehr-Gobbers, K., Storck, C.: Reputation Management. Springer, Cham (2011)
11. Rao, H.: The social construction of reputation: certification contests, legitimation, and the survival of organizations. In the American Automobile Industry: 1895–1912
12. Corradini, I., Nardelli, E.: La reputazione aziendale: aspetti sociali, di misurazione e di gestione, Franco Angeli (2015)
13. Corradini, I.: Cyber risks, social network e rischi reputazionali. In: ICT Security Conference, Rome (2014)
14. Chang, E., Dillon, T.S., Hussain, F.K.: Trust and reputation relationships in service-oriented environments. In: Third International Conference on Information Technology and Applications (ICITA 2005), Sydney, NSW, vol. 1, pp. 4–14 (2005)
15. Jøsang, A., Ismail, R., Boyd, C.: A survey of trust and reputation systems for online service provision. Decis. Support Syst. **43**(2), 618–644 (2007)
16. Rindova, P.V., Fombrun, C.J.: Constructing competitive ad-vantage: the role of firm–constituent interactions. Strateg. Manag. J. **20**(8), 691–710 (1999)
17. Feldman, P.M., Bahamonde, R.A., Bellido, I.V.: A new approach for measuring corporate reputation. RAE, Revista de Administracao de Empresas, **54**(1), 53–66 (2014)
18. C-Suite 2018, Global Business Policy Council & AT Kearney, Annual Survey of Global Business Executives (2018)
19. Corradini, I.: Building a Cybersecurity Culture in Organizations: How to Bridge People and Digital Technology. Springer, Cham (2020)
20. Ponemon Institute, 2018 Cost of a Data Breach Study: Global Overview (2018)
21. Carreras, E., Alloza, A., Carreras, A.: Corporate Reputation. LID Publishing, London (2013)
22. Fombrun, C.J., Gardberg, N.A., Sever, J.M.: The reputation QuotientSM: a multi-stakeholder measure of corporate reputation. J. Brand Manag. **7**(4), 241–255 (2000)
23. Walsh, G., Beatty, S.E., Shiu, E.M.K.: The customer-based corporate reputation scale: replication and short-form. J. Bus. Res. **62**, 924–930 (2009)
24. Hillenbrand, C., Money, K.: Corporate responsibility and corporate reputation: two separate concepts or two sides of the same coin? Corp. Reputation Rev. **10**(4), 261–277 (2007)

Author Index

I. Corradini et al. (Eds.): AHFE 2020, AISC 1219, pp. 141–142, 2020.
https://doi.org/10.1007/978-3-030-52581-1

The manufacturer's authorised representative in the EU is Springer Nature Customer Service Centre GmbH, Europaplatz 3, 69115 Heidelberg, Germany. If you have any concerns regarding our products, please contact ProductSafety@springernature.com

Printed and bound by CPI Group (UK) Ltd, Croydon, CR0 4YY
30/09/2026
02225575-0014